AF323654

THE ECONOMICS OF SOYBEAN DISEASE CONTROL

THE ECONOMICS OF SOYBEAN DISEASE CONTROL

Nicholas Kalaitzandonakes

Professor and F. Miller Chair of Innovation Studies
Director, Food Equation Institute
University of Missouri

James Kaufman

Project Director
Food Equation Institute
University of Missouri

and

Kenneth Zahringer

Senior Research Associate
Food Equation Institute
University of Missouri

CABI is a trading name of CAB International

CABI	CABI
Nosworthy Way	745 Atlantic Avenue
Wallingford	8th Floor
Oxfordshire OX10 8DE	Boston, MA 02111
UK	USA
Tel: +44 (0)1491 832111	Tel: +1 (617)682-9015
Fax: +44 (0)1491 833508	E-mail: cabi-nao@cabi.org
E-mail: info@cabi.org	
Website: www.cabi.org	

A catalogue record for this book is available from the British Library, London, UK.

ISBN-13: 9781780648088 (hardback)
         9781780648095 (ePDF)
         9781780648101 (ePub)

Commissioning Editor: David Hemming
Editorial Assistant: Tabitha Jay
Production Editor: Kate Hill

Typeset by SPi, Pondicherry, India
Printed and bound in the UK by CPI Group (UK) Ltd, Croydon, CR0 4YY

Contents

Preface

As the world population passes 7.5 billion on its way to a predicted 9.5 billion by 2050, the global agrifood industry faces the challenge of feeding more people from a fixed or even decreasing amount of agricultural land, water, and other resources. This can happen through significant productivity gains in agriculture and through reductions in postharvest losses and food waste. In crop production, developing higher yielding varieties and improved crop protection methods that reduce the amount of produce lost to weed, insect, and pathogen pests are the primary strategies for securing productivity gains. Both of these strategies are constant areas of research by universities, public research institutes, and private firms.

In this book we are concerned with one aspect of the quest for productivity, the economics of managing diseases that affect soybean production. The essential scientific research undertaken is just the first step in meeting the productivity challenge. The ultimate goal is to develop new methods and practices that are adopted by farmers and effectively used in the field. In this context, economic considerations naturally come to the fore. New methods and practices must be offered to farmers at prices they would be willing to pay and hence commensurate with the added value they derive from these innovations. Producer valuations, in turn, place limits on the investments that developers make in research and development (R&D), as they must get a sufficient return on their R&D spending in order to fund future product innovation. Thus, market feedback loops between R&D and field performance strongly influence the nature and quantity of future innovation. Our purpose here is to examine the crop protection practices that are available to soybean farmers for controlling diseases and associated yield losses, the inherent complexity in the crop protection decisions that farmers must make, the economic impacts of such decisions, and the linkages between producer decisions and the decisions made by technology developers in R&D.

We focus on diseases because they are substantially different from weed and animal pests and because they present unique management challenges that have not been researched sufficiently. Oomycetes, fungi, nematodes, viruses, and bacteria represent different classes of organisms that have little in common with each other. Control measures effective on one pathogen may have no effect on, or even promote the growth of, others. They often have a high degree of genetic diversity, high mutation rates, and sometimes multiple generations in one growing season, all of which can allow them to rapidly adapt and overcome chemical controls, genetic resistance in the crop, or agronomic control measures.

We focus on soybeans because they are one of the world's major commodity crops. Soybeans are a source of edible oil for human consumption, protein meal for livestock feed, and a growing list of specialty food and industrial products. Global soybean production has roughly doubled since the beginning of the 21st century and has increased more than sevenfold in the last 40 years. Soybean cultivation now occupies some 300 million acres of land in well over 100 countries on every continent except Antarctica, making it the world's primary oilseed crop and fourth most widely grown row crop. Beyond their native China, soybeans are grown in environments as diverse as Egypt, Indonesia, and Finland, in addition to the major producing countries of Argentina, Brazil, and the United States.

We begin this book with a historical account of soybeans and soybean production, charting their amazing growth from a simple forage crop of little consequence to one of the world's most important commodities. We go on to review the range of significant soybean diseases, the ways they damage the crop, and their impacts on soybean production. We place those impacts in context by contrasting them with production losses from weed and insect pests.

We then narrow our focus to one class of soybean pathogens for a more detailed analysis: the oomycetes, which cause both seedling and mid-season diseases. Within this context we examine the various chemical, genetic, and agronomic practices available to farmers for controlling oomycetes, and analyze their decision-making process in choosing among alternatives. Next, we analyze the economic impacts of the aggregated decisions of soybean farmers and how those impacts are distributed among producers and consumers worldwide. We also examine how producer decisions, individually and in the aggregate, along with technical, regulatory, and other factors, shape industry decisions on R&D investments and the supply of future innovations for disease control.

We focus on one class of diseases in one specific crop in order to provide detail and the essential data that is necessary for the analysis. However, our methods and conclusions are broadly applicable to all manner of disease control strategies in a wide range of crops. Professional economists will be informed by our work but we have limited the extent of technical discussion to make our work accessible to non-economists. Our intended audience, therefore, includes agronomists, crop protection specialists, industry managers and policy planners with an interest in sustainable agricultural production.

Acknowledgements

This book and its underlying research was made possible by the grant project "Integrated management of oomycete diseases of soybean and other crop plants." We would like to express our appreciation to the USDA's National Institute of Food and Agriculture for funding this project through the Agriculture and Food Research Initiative grant 2011-68004-30104.

Moreover, we would like to extend special thanks for the ongoing collaboration and friendship from all of the participants in the project, especially:

- Brett Tyler, Professor, and Director of the Center for Genome Research and Biocomputing, Oregon State University.
- John McDowell, Associate Professor, Department of Plant Pathology, Physiology and Weed Science, Virginia Polytechnic Institute and State University.
- Joel Shuman, Project Manager, Department of Plant Pathology, Physiology, and Weed Science, Virginia Polytechnic Institute and State University.
- Saghai Maroof, Professor, Department of Crop, Soil and Environmental Sciences, Virginia Polytechnic Institute and State University.
- Wayne Parrott, Professor, Department of Crop and Soil Sciences, University of Georgia.
- Paul Morris, Associate Professor, Department of Biological Sciences, Bowling Green State University, Ohio.
- Alison Robertson, Associate Professor, Department of Plant Pathology and Microbiology, Iowa State University.
- Martin Chilvers, Assistant Professor, Department of Plant, Soil and Microbial Sciences, Michigan State University.

We also thank Laura Sweets, Kenneth Sudduth, and Patricia Quackenbush for taking time to share with us their knowledge in their respective areas of expertise.

Finally, special thanks to Carl Bradley, Tom Allen, and Paul Esker for sharing their data set on US crop disease losses.

1 Soybeans: the Emergence of a Global Crop

History of the Soybean

It is well known that the soybean (*Glycine max*) was originally domesticated in China, but a more detailed picture of its history has been slow to develop. It does not appear that soybeans were among the crops raised by Neolithic Chinese farmers; that honor goes to rice and various species of millet. The domestication process for soybeans may have begun as early as 3000 BC, nearly two millennia after domesticated rice was present in archaeological data (Guo *et al.*, 2010). That process likely took 1000–2000 years, as the first known historical reference to soybeans was in the Book of Odes, dating back to 1200–1000 BC. These writings refer to soybeans brought as a tribute to the court of the Chou dynasty by recently conquered tribes from Manchuria, in the extreme north-east of modern-day China. Chou officials were so impressed with the soybean's potential as a crop that they distributed seeds throughout northern China beginning in 664 BC. Due to the absence, at the time, of archaeological evidence of comparably early soybean agriculture in other parts of China, most early writers believed that the soybean was first domesticated in or around Manchuria (Ho, 1975).

More recent research has complicated the picture. Archaeologists have discovered soybean remains in central China and in Manchuria dating from 1000 to 600 BC, consistent with the existing historical data. However, they have also uncovered inscriptions referring to soybeans from the Yin and Shang dynasties, around 1700 BC, much earlier than the references in the Book of Odes. This suggests that soybeans were known in north and central China prior to the conquest of Manchuria, and again brings to the fore the question of origins. There is still some evidence supporting the theory that domesticated soybeans originated in north-east China and Manchuria. This area does contain the most extensive population of wild members of the genus *Glycine*, but native species are widely distributed throughout the eastern two-thirds or so of China.

Manchuria is the current center of Chinese soybean production, but this may or may not give us accurate insights into the origins and travels of domestic soybeans (Qiu and Chang, 2010).

Modern archaeological, biological, and genetic investigations suggest other possibilities for the soybean's area of origin. Chinese civilization and agriculture began in the middle Yellow River valley of central China, so it is not unreasonable to investigate whether soybeans might have been domesticated in this region as well. The Shang inscriptions mentioned above indicate that the earliest soybean production took place in central China. Agricultural scientists have noted that the blooming date of domestic and wild soybeans are essentially the same around 35°N, the location of the middle reaches of the Yellow River, while the two diverge as one travels either north or south from there. The protein content of domestic soybeans is also most similar to wild types from the region of 34–35°N. These two characteristics suggest at least a close relationship between domestic soybeans and those from the Yellow River valley (Qiu and Chang, 2010). Southern China, in the Yangtze River Valley, has also been suggested as a possible area of origin. This region is heavily populated with wild soybean varieties, and many of these show the strongest short-day character, a central trait of domestic soybeans. This strongly suggests southern China as an area of origin. In addition, in-depth genetic studies have shown that domestic soybean varieties are most similar at the genomic level to wild types from the Yangtze River Valley and more southern areas (Guo *et al.*, 2010). If there is only one region of origin for domestic soybeans, this is the strongest evidence for southern China as that region. Given the mix of evidence currently available, however, the possibility of multiple origins cannot be ruled out.

Whatever its origins may have been, by the end of the 1st century AD the soybean had spread throughout China and the Korean peninsula. It took a few more centuries to reach Japan; the first historical reference to soybeans there dates to AD 712 (Hymowitz, 1990). During the timespan from the 1st century AD until the European voyages of discovery began in the 15th century, neighboring peoples took to soybean cultivation, developing locally adapted landraces in Japan, Indonesia, the Philippines, Vietnam, Malaysia, Thailand, Burma, Nepal, and northern India. Soybean cultivation spread through this area as land and sea trade routes developed and disparate groups adopted the soybean as a staple crop (Hymowitz, 1970).

Hymowitz (1990) tells us that Europeans experienced soybeans and soybean-derived food products with their visits to the Orient in the 16th and 17th centuries, but they may not have realized the connection at the time. The main foods made from soybeans, such as miso, tofu, and soy sauce, bear very little resemblance to the whole bean. At least one 17th century European visitor recorded his surprise that cheese was so popular among the Japanese (apparently a mistaken reference to tofu), yet they consumed no other dairy products. None the less, soy sauce in particular was a popular trade good exported to Europe during the 17th century. The full story was finally told to Europeans in 1712 with the publication of *Amoenitatum Exoticarum* (Exotic Novelties) by one Engelbert Kaempfer, in which he correctly described in detail both the soybean plant and the process for making soy sauce. Soybean seeds were brought to Europe some time not too long after that, as Linnaeus described the plant from specimens growing in a garden in the Netherlands in 1737. While the bulk of European soybeans during this time were grown as novelties or as botanical samples, by the early 19th century a small amount of soybeans was produced for animal feed.

The soybean was also introduced into North America during the 18th century. The earliest known importation of seeds was in 1765, by a sailor from Georgia by the name of Samuel Bowen, who had worked in the China trade. Bowen grew soybeans and produced soy sauce on his plantation outside Savannah for many years and sent seeds to the American Philosophical Society in Philadelphia for study and further distribution to other farmers (Hymowitz and Shurtleff, 2005). Another noteworthy introduction of soybeans came in 1770, when Benjamin Franklin sent soybeans from France to friends in America. Soybeans became increasingly popular, and many other individuals imported seeds to North America over subsequent years. By the 1850s, soybeans were a fairly common forage crop. By 1924, US production reached 5 million bushels per year (Hymowitz, 1990) before entering a period of growth in the 1930s. During the 1940s and 1950s, soybean production began growing rapidly in response to increasing demand as the potential uses of the crop became more apparent (Harper, 1958).

In response to the continued popularity and potential of soybeans, the US Department of Agriculture (USDA) organized two expeditions to Asia in the 1920s to collect soybean germplasm. In the second of these, two USDA scientists sent back 4451 germplasm accessions collected in China, Korea, and Japan during 1929–1931 (Hymowitz, 1984). These germplasm samples have proven to be a valuable source of genetic variation for public and private breeders to the present day, especially as a source of pest resistance. PI88788, the source of resistance to soybean cyst nematode, arguably the most serious soybean pest in North America, was among those samples collected in the 1920s (Hymowitz, 1990).

Soybeans were also brought to South America in the 19th century, being grown in Brazil as early as 1882. During the first half of the 20th century, it was a fairly common crop, especially among Japanese immigrant farmers. University research on soybeans was taking place as early as 1914. Still, soybeans made only a minor contribution to Brazilian agricultural production until the 1960s. During that decade, the Brazilian government, recognizing the burgeoning growth in world demand, designed and implemented several policy initiatives aimed at promoting soybean production and processing. Production grew slowly but steadily; by 1970, the soybean harvested area totaled 3.2 million acres. Over the succeeding 25 years, both harvested area and yield saw dramatic increases. By 1995, the harvested area had increased more than eightfold, to over 28.4 million acres, while yield nearly doubled (Warnken, 1999).

The Brazilian Agricultural Research Corporation (Embrapa), founded in 1973, was instrumental in the growth of the industry by developing new soybean varieties that could thrive in the unique soils of the *Cerrado* region (Garrett *et al.*, 2013), opening millions of acres of previously unusable land to soybean production. Soybean production in the *Cerrado* began in earnest around 1980, and accounted for nearly half of the Brazilian harvested area by 1995 (Warnken, 1999). Another round of expansion began in the early 2000s and continues to the current time; in 2016, the harvested area amounted to some 84 million acres (USDA, 2017d). Embrapa has been an integral part of a suite of government programs of agriculture research, extension, and credit availability that produced an average 1.9% annual increase in soybean yield over the period 1975–2010 (Martha *et al.*, 2012). Yield has averaged around 44.5 bushels/acre during the 2010s. The increases in area and yield have allowed Brazilian soybean production to grow at an annual pace of 6.2% since 1977 (USDA, 2017d).

Brazil is not alone in South America; other countries have also made concerted efforts to promote soybean production and have been quite successful. In Argentina, the harvested area, which stood at 3.1 million acres in 1977, increased to 48.2 million acres by 2016, with yields approaching the 45 bushels/acre mark. The production area in Paraguay was less than 250,000 acres in 1970 but increased to 8.4 million acres in 2016. Both Bolivia and Uruguay had negligible soybean production in 1970 and have since increased their harvested areas by around 1000-fold, to over 2.5 million acres each (USDA, 2017d).

Soybean Production and Distribution

Soybean production

Over the last 60 years, soybeans have become one of the major agricultural commodities traded on global markets. Since 1964, when USDA record keeping on soybean production began, world production has grown 4.8% per year on average, a pace that resulted in a doubling of production every 14 years. This prodigious growth was largely the result of the greatly increased production in the USA, Argentina, and Brazil. Indeed, currently, Argentina, Brazil, and the USA account for over 80% of global soybean production (USDA, 2017d), and world soybean production in 2017 was 12.9 billion bushels. Figure 1.1 shows the trend of increasing production for the major soybean-producing countries.

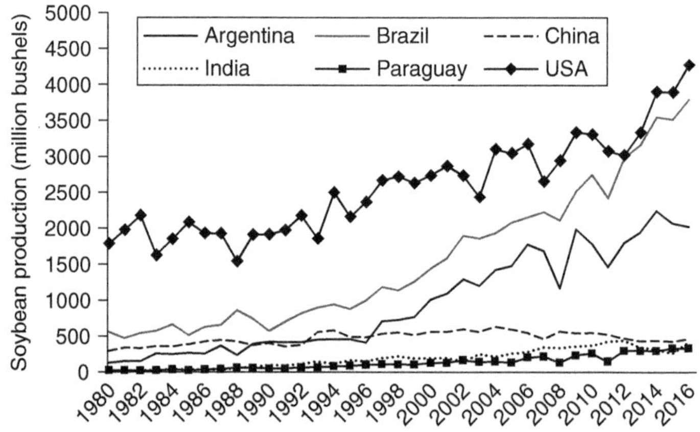

Fig. 1.1. Soybean production of the top six soybean-producing countries. (Data from USDA, 2017d.)

This explosive increase in production has been a result of both increased area under cultivation and increased yield per acre. Since 1964, the global land area devoted to soybean production has increased by 2.7% per year, to just over 295 million acres in 2017. Brazil's contribution to the harvested area is unique in that it came from the previously uncultivated land of the *Cerrado* region. Increases in soybean area in most other countries, including Argentina, India, and the USA, have come largely at the expense of other crops, as they had relatively little unused potential agricultural land. The US soybean area has changed very little since 1980; production gains have come primarily through yield increases (USDA, 2017d).

The USA has long been the world's largest producer, although its market share has decreased steadily with the growth of production in Brazil and Argentina, especially since about 1995. Currently, the USA produces roughly 34% of the world's soybeans, followed closely by Brazil with 30%. Argentina held approximately an 18% share for 10 years or so before dropping to 16.5% in 2017. The remainder is split among several smaller producing countries. These relative production levels have been changing continuously over the years, and likely will continue to do so. Figure 1.2 shows how the market shares of the top producers have changed since 1980. The variation in market shares of these three countries is a result of differences in their rates of production growth. Most countries with lower production quantities have not significantly increased production levels in several years, hence their small and steadily declining market shares (USDA, 2017d).

Soybean products and exports

Soybeans and other oilseeds are different from most other major agricultural crops in that they are transformed into two separate products before being used by consumers. Most soybean production goes to crushing, where soybean oil

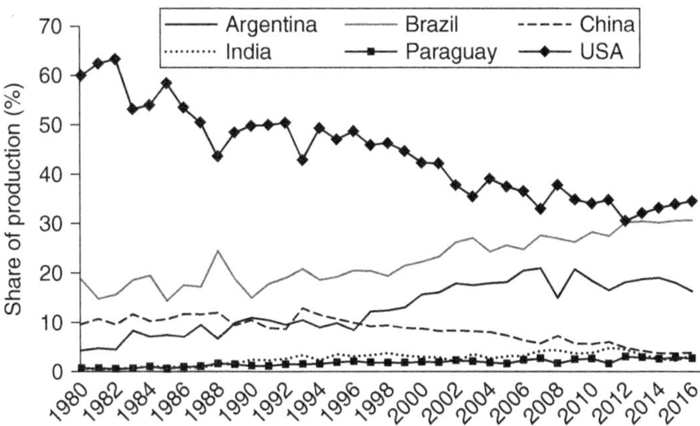

Fig. 1.2. Share of world soybean production of the top six soybean-producing countries. (Data from USDA, 2017d.)

is separated from the dry seed matter, or meal. Both of these products have various feed, food, and industrial uses. Nearly all soybean meal is used as animal feed, while the oil is predominantly used for human food consumption. The top soybean producers differ greatly in the proportion of their production that is exported as whole beans, meal, or oil. Although Argentina sits in a somewhat distant third place in soybean production, it is far and away the world leader in exports of both soybean meal and oil (Figs 1.3 and 1.4). This is primarily due to an export duty regime that gives preferential tax treatment to soybean products over whole beans (USDA, 2017a). Overall, 71% of Argentine soybean production is crushed domestically, with 95% of the resulting meal and 79% of the oil being exported to world markets. Brazil and the USA each crush about 44% of

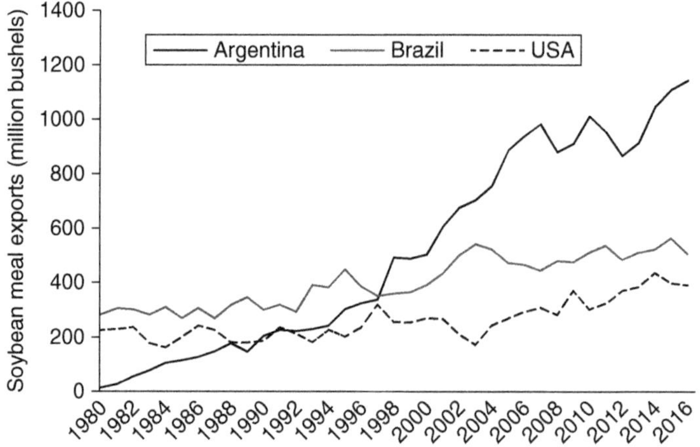

Fig. 1.3. Soybean meal exports production of the top three soybean meal-producing countries. (Data from USDA, 2017d.)

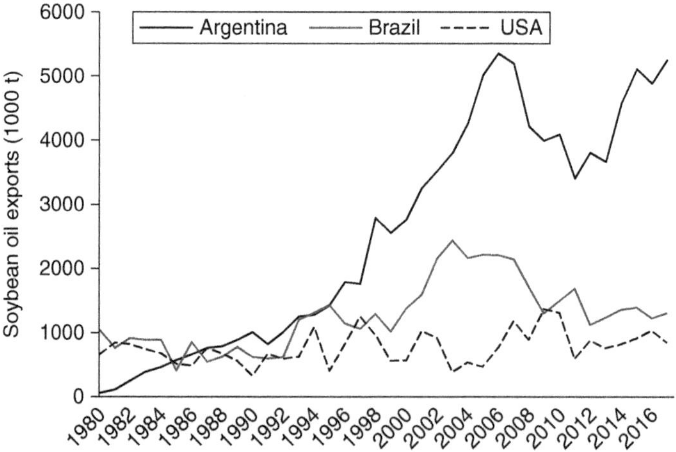

Fig. 1.4. Soybean oil exports production of the top three soybean oil-producing countries. (Data from USDA, 2017d.)

their production, although Brazil exports a much greater proportion of the meal and oil than the USA, where more goes to domestic feed and food use. Since 2012, Brazil has moved into the role of the world's largest exporter of whole soybeans, with the USA a close second (Fig. 1.5).

Soybean consumption and imports

Most key soybean-producing countries are also among the major consumers of soybean products, and nearly all of that demand is filled by domestic production. Some importing countries are also among the largest consumers. As Fig. 1.6 shows, China is the largest and fastest-growing importer of whole

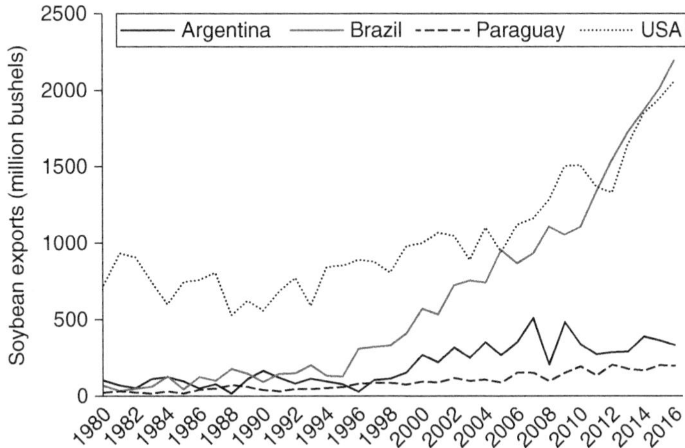

Fig. 1.5. Whole soybean exports of the top four whole soybean-producing countries. (Data from USDA, 2017d.)

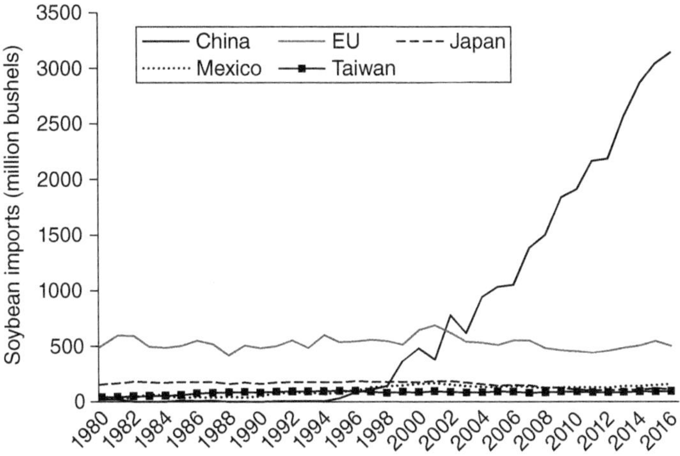

Fig. 1.6. Soybean imports of the top five soybean-importing countries. (Data from USDA, 2017d.)

soybeans, its consumption having increased by more than 10% annually since the early 1990s. China is responsible for most of the increase in global soybean demand since that time. Some 64% of soybeans imported into or produced in China are crushed, and 97% of all Chinese soybean products are used domestically. This dramatic increase in demand for soybeans has been driven mainly by increased meat consumption, primarily of poultry and pork. Global pork demand has been quite strong, growing at 1.9% per year over the past 20 years. Much of this growth in demand has occurred in China (Fig. 1.7); pork consumption there increased at a 2.3% annual rate during 1995–2015, compared with a 1.6% per year increase in the rest of the world.

The strongest growth in demand of all meats has occurred in the poultry industry. Not only has the growth in consumption been considerably larger than other types of meat, at 3.6% per year from 1995 to 2015, but, as Fig. 1.8 shows,

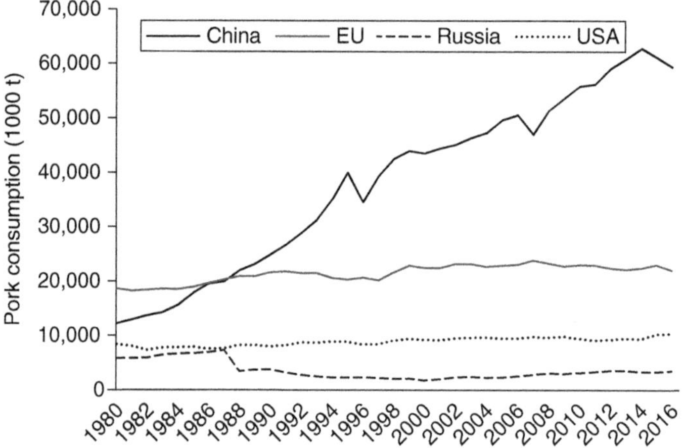

Fig. 1.7. Pork consumption of the top four pork-consuming countries. (Data from USDA, 2017d.)

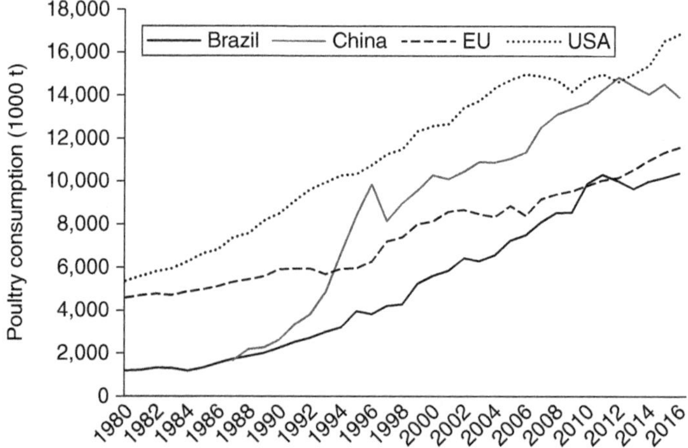

Fig. 1.8. Poultry consumption of the top four poultry-consuming countries. (Data from USDA, 2017d.)

it has been much more broad-based. All major poultry-consuming countries have shown a consistent growth in demand for many years. This tremendous increase in demand for animal products, in addition to a trend toward more balanced feed formulations that include increased oilseed ingredients in addition to grains, has translated into dramatically higher soybean meal demand. China's feed industry maintains sufficient crushing capacity to meet domestic feed demand; although China is the world's largest whole soybean importer, they import almost no soybean meal (USDA, 2017d).

In addition to China, the USA and European Union (EU) are also major soybean meal consumers; together, these three account for well over half of world consumption. Each has its own strategy for filling that demand. The USA is self-sufficient in both production and crushing capacity. The EU, by contrast, is the world's major importer of meal, importing over 60% of its consumption and buying 30% of the soybean meal traded on global markets (USDA, 2017d).

Brazil, China, the EU, and the USA are major consumers of soybean oil as well as meal. All of these countries, however, currently fill most of their consumption from domestic crushing capability. China has been a major oil importer, but its imports have been on a downward trend since the early 2000s. The EU's comparatively modest crushing capacity is sufficient for its oil demand, while its meal imports reflect a disproportionately large feed demand. India has recently emerged as the leading soybean oil importer. Rising family incomes and increasing consumer health consciousness are driving steady, large increases in demand for all vegetable oils (USDA, 2017b). Over the past few years, India has purchased over one-third of the soybean oil traded on world markets (USDA, 2017d).

Future Growth in Soybean Demand and Supply

Several factors will influence the future of demand for soybeans and soybean products. At the center of the constellation is world population growth, which by itself will bring about increased food demand. Population growth has been slowing for many years, but it is expected to remain in the 0.5–1.0% per year range through to 2050, with the world population topping 9.6 billion by that time. However, this growth is not distributed evenly across all countries and regions of the world. As of 2010, 75 countries had below-replacement fertility rates, including 43 developed countries. The developed world, as a whole, is expected to be in population decline during 2050–2100, with North America being a notable exception to this trend. Most of the developing world will maintain higher growth rates in the future; some countries in this group may see their population triple by 2100. Sub-Saharan Africa in particular continues to be a center of population growth, with an expected rate of around 2% per year during 2030–2050 and 1% per year from 2050 to 2100. By 2050, five of the 20 most populous countries will be least-developed countries (UN, 2015). Simply feeding the increased number of people on the planet will require growth in agricultural production, although the slowing rate of growth will put less pressure on the agriculture sector than has been the case to the present day.

Increasing per capita income will interact with population growth patterns to influence overall food demand. Food products have a positive income elasticity of demand, meaning that consumption increases with income, but only up to a point. Beyond about 2700 kcal per day per person, food demand grows much less with income gains, and above around 3000 kcal per day per person, it increases very little or not at all. In the developed world and some of the more populous developing countries, notably China, per capita consumption is already at or near these levels, which will moderate future demand growth. Since most population growth will take place in the developing world, where the potential for increasing per capita income is also the greatest, potential future food demand could increase substantially among these countries. Whether this potential is translated into higher effective demand is dependent on how much per capita income actually increases. To the extent that individuals are unable to rise out of poverty, future food demand will grow less. Moreover, this relationship is not constant across all cultural groups. India, for example, has experienced little or no growth in food demand over the last 25 years, despite a significant increase in per capita income. If this condition changes in the future, it could make a noticeable difference in world food demand, given that India will soon surpass China as the world's most populous country (Alexandratos and Bruinsma, 2012).

Continued economic growth will also change the structure of food demand. As population and household incomes both rise, especially in developing countries, not only does overall food demand increase, but demand also shifts toward higher-quality animal protein foods. This trend has been under way for some time in many parts of the world, and increases in meat consumption are expected to continue for some time. This will ensure that feed demand will also continue to increase in all sectors of livestock production and, in turn, that demand for soybeans as a major feed ingredient will continue to grow as well.

Increasing meat consumption is a major factor driving soybean demand, but it is not the only factor. The protein consumption dynamics or PCD model estimates total human dietary protein content based on population and per capita income. Using this model, Sonka *et al.* (2004) projected that vegetable protein consumption would increase by 38% and animal protein consumption by 81% during 2001–2025, which comes out to 1.3% and 2.4% per year, respectively. Since soybeans are traditionally a major source of vegetable protein, especially in Asia, a significant proportion of that increased consumption should translate into soybean demand. Soybeans are also a major source of oil for human consumption. Global soybean oil consumption has increased at a 4.6% annual pace over 1995–2015, a trend that should also continue with an increasing population (USDA, 2017d).

Recent genetic research efforts are developing new soybean varieties with specialized oil composition, some producing "healthy" oils for human consumption and others oils suitable for a variety of industrial uses (USDA, 2017d; USSEC, 2015). In fact, a variety of smaller-scale industrial uses of soybean oil are already included in total consumption, including inks, paints, detergents, and lubricants. One such industrial use of soybean oil that could become significant in the future is energy production. Soybean oil is a major feedstock for biodiesel fuel production. Since this application is often based on government mandates, it is somewhat difficult to project future trends. However, there is ample potential for this to become a significant category of soybean use that could play a growing role in overall demand (Alexandratos and Bruinsma, 2012).

For producers to be able to meet such future demand for soybeans and processed products, they will have to implement a multi-pronged approach, potentially including increases in cultivated area, growth in actual and attainable yield per unit area, and a reduction in production losses from pests. A variety of pests can inhabit soybean fields and reduce yield below what it otherwise would be. Weeds, animal pests (arthropods, nematodes, and others), and pathogens (fungi, bacteria, and viruses) all take a toll on soybean production. The focus of this book is on soybean losses from pathogens and the economic decisions to control them.

Putting Disease Control in Context

Severe crop disease outbreaks have sometimes become infamous because of their unexpected nature and catastrophic impacts. Around 1840, a new strain of the oomycete *Phytophthora infestans* arrived in Europe by an unknown route and proved to be especially virulent to the potato varieties then cultivated in Europe. People in Ireland were heavily dependent on the potato as a subsistence crop, planting extensive monocultures of susceptible plants. By the time the epiphytotic had run its course, some 1.5 million Irish had died of starvation and a similar number had emigrated to the Americas (Fry and Goodwin, 1997b). In 1942, an extended period of unusually favorable weather conditions in the state of Bengal in eastern India fostered an abnormally large outbreak of the fungus *Cochliobolus miyabeanus*, causing losses of 50–90% of the rice crop that year. With a population almost completely dependent on rice for subsistence, an estimated 2 million people died of starvation in the ensuing famine of 1943 (Padmanabhan, 1973).

Events such as these are rendered all the more dramatic because they are rare. However, the pattern of a disease seemingly coming out of nowhere, laying waste to a crop, and then disappearing the next year is repeated on a smaller scale in many agricultural fields somewhere every single year. As Strange and Scott (2005) pointed out, the unpredictable nature of disease incidence and severity may be their most distinguishing feature and an inherent obstacle in their management.

When economic considerations are included, disease management becomes still more complex. For any pest causing crop loss, there is a damage threshold below which it is not economical to attempt control. At this economic threshold, the expected economic gain from the extra yield is just equal to the cost of control (Nutter *et al.*, 1993). Zero crop damage is almost never economically optimal. Farmers, then, must ascertain which disease(s) are likely to be present in a particular field and develop expectations about which of these might be severe enough to warrant expenditures on control effort. How farmers form expectations about the incidence and severity of disease, how they choose among disease control practices, and how they decide when to implement them are therefore key matters in the economics of soybean disease management. We address these issues in the following chapters, beginning with an overview of the nature, incidence, and severity of the major diseases afflicting soybean production.

Summary

While the soybean has a long history in world agriculture, it has only recently joined the ranks of the most widely grown crops. Since the middle of the last century, soybeans have become the world's premier oilseed crop and a key resource for meeting the future food needs of a growing population. As populations and per capita incomes increase across the developing world, demand for food, especially for higher-quality animal protein, is also expected to increase. Farmers will need to employ a variety of tactics to meet this growing demand, one of which is to decrease yield losses to disease and other pests. In this book, we examine the economics of disease control in soybean production. In particular, we analyze how farmers decide whether to expend resources to protect their soybeans from disease, how farmers choose among alternative disease control methods, how farmers decide whether to adopt new disease control practices, how farmers' decisions are linked with the research and development (R&D) strategies of upstream industries that determine the future solutions for disease control and associated losses, and, finally, how consumers and societies are affected by such decisions and strategies.

2 Soybean Disease and Production Losses

Organisms in nature are engaged in a constant struggle for survival against other competing organisms, and crop plants are no exception. Many different organisms have adapted to the human-created environment of agricultural cropland, either competing with crop plants for resources or preying on them, and thus interfering with the crop producer's plans. To the extent that these other organisms are successful in their quest for survival, they partly or completely compromise the growth and productivity of crops and therefore are regarded as pests. Some pests are microorganisms that physically invade crop plant tissues and disrupt the metabolic functions of plant cells, sometimes destroying them in the process. These pathogens are mainly nematodes, fungi, and oomycetes, along with some bacteria and viruses. Other, larger organisms attack crops from the surface and consume plant tissues as food. This group of animal pests consists primarily of insects and other arthropods but also includes some reptiles and mammals. Finally, some plants have independently adapted to the environment of agricultural fields, without the assistance or encouragement of humans. Weeds compete directly with crops for sunlight, water, and soil nutrients. The different types of pest organisms can interact in various ways. Weeds, for instance, can serve as alternative hosts for pathogens and insects awaiting the arrival of the next crop, while insects are often vectors for viral and bacterial diseases.

Measuring the amount of production given up to the various pests is no mean feat, for a few reasons. Chief among them is that the counterfactual condition – what the yield would be in the absence of pest losses – is unknown and must be estimated. Moreover, the proper value of this estimate is peculiar to the region, the cultivar, the prevailing climatic conditions, and possibly the field in question. Using the maximum attainable yield for a particular cultivar is not the proper basis for comparison, as this measure assumes optimal environmental conditions, including a lack of significant abiotic stress from water availability, soil nutrients, and the like. Such ideal conditions are by no means guaranteed to exist in any given field, and it may be prohibitively expensive to approximate them. Instead, the actual yield is often compared to an estimate of the attainable yield for that area, which assumes only that pests are controlled effectively and cause no damage, while accepting existing abiotic stressors as given (Nutter *et al.*, 1993). Attainable yield can be estimated based on actual yields across a region using any of several types of model of varying complexity (e.g. Naab *et al.*, 2004; Aggarwal *et al.*, 2006; Affholder *et al.*, 2013). However, developing a reliable method of assessing the accuracy of such estimates is problematic, as the true value of attainable yield remains unknown and can change from year to year as rainfall, temperature patterns, and other environmental stressors change.

Measurement difficulties notwithstanding, there have been comprehensive attempts at estimating potential and actual losses to pests in soybeans as well as other crops. Such studies strongly suggest that pests can cause a prodigious amount of damage to food crop production. In the 1960s and 1970s, it was estimated that over one-third of world food production was consumed or destroyed by pests of various kinds (Zimdahl, 2007). More recent and detailed estimates indicate that the situation has changed only slightly. Potential losses to pests – the increase that would be experienced in the absence of pest management efforts – range from 50% or less in barley and wheat to more than 80% in sugarbeet and cotton. Estimates of actual pest losses in various crops range from 25% to 40% (Oerke and Dehne, 2004; Oerke, 2006). All of these are estimates of preharvest losses; losses in storage are additional. Crop protection is the quest to minimize the damage caused by pests of all sorts in the most cost-effective manner, with the goal of maximizing the economic value of production in agricultural fields. In order to gain some perspective on the relative magnitude of crop loss due to disease, we first take a brief look at the impact of weeds and animal pests on soybean production.

Production Losses to Weeds

In a widely used textbook, Zimdahl (2007) notes that, while every individual weed scientist, gardener, and homeowner has a very clear idea of what weeds are, there is no single definition of what constitutes a weed that is accepted by the entire scientific community. What is widely accepted is that the definition does not hinge on any objective characteristic of the plant in question. Some plants regarded as weeds produce flowers, or can be important in erosion control, or are edible by humans or livestock, or have valuable medicinal properties. Others are parasites, and some are dangerously toxic to humans or livestock. Some modern-day weeds, such as *Chenopodium* spp. or *Amaranthus* spp., were important crops for indigenous peoples, providing a significant portion of their diet. Most common definitions are admittedly anthropocentric, revolving around the observation that a weed is a plant growing in a location where humans do not want it to be and is interfering with human plans for that location.

Only a small proportion of the more than 350,000 plant species inhabiting the Earth are relevant to agriculture, whether as crops or as weeds. No more than a few hundred plant species have ever been domesticated, and around 15 presently provide the bulk of the human food supply. The group of most important weeds comprises some 200 species, nearly half of which reside in only three families: *Poaceae*, the grasses; *Asteraceae*, the sunflower family; and *Cyperaceae*, the sedges. All members of this group of important weeds share a few characteristics that contribute to their unique ability to compete effectively with crops and thus present a continuing problem for producers (Zimdahl, 2007):

1. They have adapted to growing in disturbed soil and are able to compete effectively in that environment. This may seem trite, but it is an important trait that sets both weeds and crops apart from the vast majority of other plants.
2. Their seeds can survive for long periods, even several years, in the soil and require no special conditions for germination.
3. They are able to emerge quickly, grow rapidly in their early stages, and produce seeds while the plant is still quite small.

With their ability to compete effectively in an agricultural environment, weeds can displace crop plants, stealing soil, light, and water resources and inhibiting the growth and productivity of crops. Weeds can thus impose considerable costs on producers in terms of both lost production and resources expended on control. Indeed, weeds are generally regarded as having the greatest potential for causing losses in soybean, as well as many other crops, of the three classes of crop pests. In the USA, studies from the 1970s onward estimate potential soybean losses from weeds at 25–50%, with some smaller-scale field experiments showing potential losses of 60–80%. Effective weed control practices can reduce these potential losses significantly; actual soybean losses to weeds tend to be in the 10–20% range, and even lower in some locations and years (Oerke, 1999).

Even at this lower level, the monetary cost of weeds can be considerable. One estimate put the annual cost of US weed losses in a selection of 46 crops, using best weed management practices, at well over $15 billion during 1989–1991 (Zimdahl, 2007). The cost of weed losses in more recent years would almost certainly be significantly higher. Weeds have proven to be at least as serious in South America, with estimated potential losses in Brazil of some 65%. In Argentina, different weed types have been known to combine to cause a complete loss of the soybean crop in some fields in the absence of weed control measures. Actual losses with weed management practices are likely around 15%. In East Asia, including China, soybean farmers employ a wide variety of weed control techniques, including hand weeding. Potential losses there have been estimated at 40–45%, with actual losses around 13% (Oerke, 1999).

Soybean weed management entered a new era in 1996 with the release of the first herbicide-tolerant (HT) biotech soybean varieties. The first associated herbicide, glyphosate, offered the advantages of a broad-spectrum herbicide that was also less expensive, less toxic, and more readily degradable in the soil. Other similar duets of HT soybeans and herbicides have since also been marketed. When producing HT soybeans, one or two applications of the broad-spectrum herbicide can replace several applications of more selective herbicides. HT soybeans were adopted very quickly and extensively in all major soybean-producing countries; in 2016, 94% of US soybean acreage was planted with HT beans. Other countries have seen even greater adoption; 96.5% of soybean acres in Brazil and 100% in Argentina were planted with cultivars that included an HT trait (ISAAA, 2016). The resulting decrease in cost and increase in effectiveness of weed control allowed strong expansion of soybean production into previously inhospitable locations, especially in South America. Since the introduction of HT technology, the soybean area increased by approximately one-third in the USA and roughly tripled in Brazil and Argentina. The dramatic increase in supply produced in these countries kept world soybean prices in check and satisfied steep increases in demand over this period (Alston *et al.*, 2014). Glyphosate became the herbicide of choice for soybean producers; in 2006, US farmers applied glyphosate alone to two-thirds of soybean acreage (NASS, 2012). However, the exclusive and intensive use of glyphosate on such a large proportion of soybean acreage placed strong selective pressure on a number of weed species. Naturally occurring HT weeds have begun to complicate weed management and increase production costs for producers, especially in Argentina, Brazil, and the USA. In 2012, US farmers reported some level of glyphosate-tolerant weeds on 44% of soybean acres (NASS, 2012). In Argentina, farmers have reported increased weed control costs of $18–121/ha, while in Brazil costs have increased by an average of $35/ha (REM, 2014). Biotechnology researchers have begun to develop new soybean varieties tolerant to multiple herbicides in order to give farmers more weed control options against herbicide-resistant weeds.

Production Losses to Animal Pests

A wide variety of animals can function as agricultural pests. Nevertheless, it is worth noting that, while some classes of vertebrates are agricultural or horticultural pests, the amount of damage they cause is insignificant when compared with that of arthropods, primarily insects, especially in row crop production. Similarly, nematodes are, strictly speaking, animals, but the manner in which they damage a plant is fundamentally different from that of insects. Insects attack plants externally to feed on them, consuming plant tissues or juices. Nematodes, in contrast, enter plant roots and feed from the inside. In the course of feeding, they directly disrupt plant metabolic processes, much as pathogens do. While some yield loss studies group nematodes with animal pests (e.g. Oerke, 1999), or treat them as a separate category (e.g. Wrather and Koenning, 2006), most authors categorize nematodes as plant pathogens (Hartman *et al.*, 1999; Agrios, 2005; Schumann and D'Arcy, 2010). We will follow this last convention here. Thus, in our brief discussion of animal pests in soybean production, we will restrict ourselves to insect pests only.

Insects are arguably the most successful class of organisms on Earth. They can be found in nearly every habitable location apart from the deepest oceans, and in staggeringly large numbers. Just over half of all species of organisms of all types in the world are insects. Insects are an integral part of the environment wherever they are found. They provide an abundant food source for a wide variety of organisms, ranging from other insects to many different vertebrates, including humans. Some play a critical role in the life cycle of many plants, transferring pollen from one to another. Scavenger insects serve as the first step in decomposing organic matter from dead animals and plants. Predatory insects prey on other insects, helping to keep their numbers in check. The scope of ecological niches filled by insects is nearly as great as the number of insect species (Pedigo and Rice, 2009).

Not all insect activities are quite as benign as the foregoing examples, however, from a human perspective; many insects are classed as pests in certain contexts. As with plant pests (weeds), the definition of an insect pest is necessarily anthropocentric and circumstantial. An insect pest is one whose activities interfere with human plans in a particular situation. Very few insects fall into this category. Of the estimated 5.5 million insect species, probably less than 1% can be considered pests. The range of significant insect pests can be even narrower in any given region. For instance, around 600 species inflict significant damage in the USA. Only some of these are agricultural pests in row crops; others damage forest or ornamental plants, spread disease or injury to humans or domestic animals, or damage stored products or possessions (Pedigo and Rice, 2009).

Although insects are a critical concern in some crops, they are not often considered a serious threat to soybean production. As late as 1986, only 4% of soybean-planted area in the USA was treated with insecticides. During the period 2001–2003, Oerke (2006) reported actual worldwide soybean losses to animal pests of 8.8% of potential production, but this included losses from nematodes. In the USA, over the same time frame, damage estimates from Wrather and Koenning (2009) indicated that soybean cyst nematode was responsible for a 4.2% annual loss of soybean production, nearly half of the loss estimated by Oerke (2006). Given the status of cyst and root-knot nematodes as significant agents of soybean production loss worldwide, it is not unreasonable to expect the US ratio to be representative. This would imply a 4–5% global production loss to insects, the lowest of any class of soybean pest.

Much of the reason for this low level of damage lies with the impressive ability of individual soybean plants as well as plant communities to repair, recuperate, and adapt to considerable amounts of damage from insect feeding. Some studies have shown no significant yield reductions from a loss of one-third of foliage, if the damage happened before or during blooming (Turnipseed and Kogan, 1976). In addition, soybean may still be experiencing somewhat of a "honeymoon" period with regard to insect pests, although it could be in its later stages. Few native Asian insects that feed on soybean have been introduced to other areas, and many native species are still adapting to the new food source (Kogan and Turnipseed, 1987). Yield losses to insects may increase in the future as the adaptation process continues.

Although overall insect damage to global soybean production may be light, insects are still significant pests in some regions. In parts of the USA, especially along the Atlantic coast, the corn earworm, *Heliothis zea*, has broadened its host base to include soybean. This seems to have happened partly as a result of changes in maize agriculture practices that have made maize a less attractive host (Kogan and Turnipseed, 1987). *H. zea* was considered the most serious insect pest of soybeans until it was surpassed by the soybean aphid, *Aphis glycines*. This native of Asia was first detected in the USA in 2000 and has since spread throughout the north-central and north-eastern USA and southern Canada. While natural parasitoids and predators largely keep *A. glycines* in check in its native range, soybean farmers and the natural environment in North America are still adapting to this exotic pest. Its relative intolerance of high summer temperatures seems to have inhibited its spread into southern soybean-producing areas (Ragsdale *et al.*, 2011). In tropical and subtropical soybean-producing regions, including the southern USA and South America, as well as parts of Asia, stinkbugs are the most serious insect pests in soybean. Only a few of these members of the family *Pentatomidae* are significant soybean pests, but these few have been known to cause large production losses when their numbers are not well controlled (Corrêa-Ferreira and De Azevedo, 2002; Leskey *et al.*, 2012).

In most areas, including the USA, soybean insect pests have proven to be controlled most effectively by integrated pest management practices, including host plant resistance, the use of natural enemies of pest insects, changes in cropping practices, and some use of insecticides. In certain soybean-growing regions, most notably South America, such practices have proven to be insufficient. In order to provide adequate insect control, some seed firms have introduced biotech soybean cultivars with insect resistance traits for South American producers. In 2016, some 60% of soybean area in Brazil and 13% in Argentina were planted with insect resistance/HT cultivars, the remainder being planted with varieties containing only HT traits (ISAAA, 2016).

Soybean Disease Pathogens

By far the most common plant disease pathogens are fungi and oomycetes (water molds). Although now considered phylogenetically distinct, oomycetes were once classed as fungi and are often grouped with them in discussions of plant pathology. Symptoms, disease progression, and treatments are sufficiently similar that the two types of microorganism can be treated as a single group for the present purposes (Sinclair and Hartman, 1999b; Grau *et al.*, 2004). Nematodes are the other common type of soybean pathogen described here. Unique among soybean pathogenic microorganisms, nematodes are animals. Even though there are only a few species that cause significant losses to soybean producers, those that do are among the major soybean diseases worldwide (Noel, 1999; Niblack *et al.*, 2004). Many viruses and bacteria are also pathogenic to soybeans. While some can be locally significant, none takes a large toll globally, regionally, or nationally among the main soybean-producing countries (Sinclair, 1999; Tolin, 1999; Tolin and Lacy, 2004). For this reason, most of our discussion here is about fungi, oomycetes, and nematodes, and their associated impacts.

Diseases differ in their symptoms and progression based on the infecting pathogen and the growth stage of the plant at the time of infection. Some oomycetes, primarily *Pythium* spp., and fungi, usually *Rhizoctonia* spp., that reside in the soil and thrive in cooler temperatures can attack soybean seeds and seedlings virtually from the moment of germination. These cause seed rots and seedling diseases that affect the plant up through the unifoliate leaf stages. Later on, other soil-borne oomycetes and fungi that prefer warmer temperatures, including *Phytophthora sojae*, can cause mid-season root and stem rots. Also during this time, airborne fungi that attack foliage, such as *Phakopsora pachyrhizi*, which causes rust, become prevalent. Nematodes tend to attack more developed root systems in the mid- to late-season timeframe and are often asymptomatic except for yield loss.

Seedling diseases

A variety of species of fungi and oomycetes can infect soybean seedlings. The most common, and most damaging, seedling disease pathogens are oomycetes from the genera *Phytophthora* and *Pythium*. Most *Pythium* spp. germinate at cooler soil temperatures, while *Phytophthora* spp. prefer a somewhat warmer environment. This difference is reflected in regional variation in the prevalence of particular species. All oomycetes require waterlogged soil for germination and infection, a condition that often pertains in early spring. Some fungi can also cause seedling disease. The most common of these is *Rhizoctonia solani*, which also germinates in cooler soil but prefers moderate soil moisture. Some species of the *Fusarium* fungus have also been associated with seedling disease. These pathogens are known to infect seedlings both individually and in concert. The most common symptom is seedling death, known as damping off, either before or after emergence. Infected seedlings can also show general root decay, as well as lesions on roots, stem, or cotyledons (Agrios, 2005).

Although fungi and oomycetes require very different soil moisture conditions for germination and infection, they are quite similar in the general aspects of their life cycle. They germinate in the spring as the soil begins to warm and move through the soil to attach to and infect plant roots. They produce filaments that penetrate the roots to extract plant nutrients. All these pathogens can survive the winter either by consuming dead crop residue or by forming structures that lie dormant in the soil, sometimes for many years, before germinating and beginning another life cycle.

While their life cycles are broadly similar, the physiological processes of fungi and oomycetes are fundamentally different. Thus, control measures must be specific to each type of pathogen, and often to particular species. Genetic resistance is often bred into elite soybean varieties, although broad-spectrum resistance is generally only partially effective, and complete resistance is usually specific to particular subpopulations of each pathogen. Chemical fungicides can be effective but also relatively costly. Due to their physiological differences, fungi and oomycetes as a rule are susceptible to different chemical toxins, although fungicides effective against both do exist, further complicating management practices.

Root and stem diseases

Nematodes

The soybean cyst nematode (SCN), *Heterodera glycines*, has been known in Asia for hundreds, perhaps thousands, of years. It is the major soybean pathogen in the USA, causing more production loss than any other, and is also a major problem in China and elsewhere. One other nematode, the root-knot nematode (*Meloidogyne* spp.), is a significant pathogen globally but does not generally cause losses as large as SCN. These and other plant-parasitic nematodes are endemic to soybean fields worldwide. These two species are obligate, host-specific parasites; their life cycles require them to infect soybean plants in order to survive and reproduce. They have very limited mobility in soil but can be carried long distances by wind, water, and animals, as well as by machinery and vehicles carrying infected soil.

Newly hatched nematodes penetrate the young roots and migrate to vascular tissue where they feed on plant juices. Their feeding activity induces cellular changes in either the root cortex or the vascular tissue that further impede nutrient flow to the rest of the plant. As their development proceeds, the bodies of the females grow and swell until they rupture the outer root covering and are exposed. Mature males exit the root for mating. Females deposit some eggs in a gelatinous mass on the root surface and retain others inside their body. When the female dies, her body wall becomes an encapsulating cyst within which the eggs can survive for an extended period of time, perhaps 12 years or more. Eggs hatch in response to favorable environmental conditions and the cycle begins anew.

Nematode infections are typically asymptomatic, making diagnosis extremely difficult. Yield losses of 20% or more have occurred with no observable symptoms. Extreme infestations can result in visible stunting and chlorosis, symptoms that are similar to those of many fungal diseases. Visible symptoms generally indicate a problem that has been moderate to severe for several years and has already resulted in significant loss of production. Nematodes may also interact with other infections or plant stressors and worsen their impacts. Recent research has indicated that SCN can make soybeans more susceptible to sudden death syndrome, described below (Niblack *et al.*, 2004).

Because nematode symptoms are rare, plant pathologists recommend periodic soil sampling and testing to determine the presence and identity of nematode species and races. In practical terms, it is impossible to eradicate nematodes once they are established in a field, so control is the only realistic option (Riggs and Niblack, 1999). Rotation to a non-host crop is widely regarded as an effective management practice if done for multiple years. However, farm profitability and crop management considerations place constraints on both the choice of non-host crop and the time period for producing crops other than soybeans. Planting cultivars resistant to nematode infection can maintain production in years when soybeans are grown (Niblack *et al.*, 2004). Resistant varieties may also control or reduce nematode populations. Measures to decrease plant stress, including weed control and ensuring sufficient soil moisture and fertility, can also help limit the effects of nematode infection. An integrated program of all three management practices is usually recommended as the most effective means of dealing with nematode problems (Noel, 1999).

Root and stem rots and similar diseases

Phytophthora sojae is the most significant oomycete pathogen in US soybean fields. It is found in numerous soybean-producing countries, although it generally takes less of a toll on production outside the USA. *P. sojae* can cause symptoms in soybeans at any stage of development. *P. sojae* can survive in a dormant state in the form of oospores in crop residues or soil for many years. When soil temperatures rise in the spring, the oospores germinate and produce sporangia, the bodies that can release zoospores, the mobile cells that can travel to and infect plant roots. The sporangia will release zoospores only when the soil is saturated or flooded, as zoospores are motile only in water by means of flagella. The zoospores attach to soybean roots and grow hyphae, the filamentous growth characteristic of fungal mold. The hyphae grow throughout the plant, slowly consuming it and producing root rot, stem lesions, stunted plants, and chlorosis in the leaves (Schmitthenner, 1999).

Other pathogens that cause seedling disease can also cause root and stem rot in later soybean growth stages. In particular, *Pythium* spp. and *R. solani* produce virtually identical symptoms by similar disease progressions and can be locally or regionally significant diseases in diverse soybean-producing areas globally (Yang, 1999a). Management practices for all of these pathogens are similar. Improving soil drainage is an essential control measure, as water-saturated soil is required for part of the oomycete life cycle. Planting resistant cultivars is also recommended to control *P. sojae* and *R. solani*, although resistant varieties are not available for *Pythium* spp. infection. Cultivation can break up mycelia in the soil and bury them deeper, delaying germination. However, this may conflict with other agronomic considerations such as the producer's choice between conventional and reduced tillage (Grau *et al.*, 2004).

Charcoal rot

Charcoal rot is present in soybean-producing areas throughout the world but is a more severe problem in the tropics and subtropics. It is caused by the fungus *Macrophomina phaseolina*, which can also infect maize, lucerne, grain sorghum, cotton, groundnut, and other field crops. Investigators have found differences in varieties of *M. phaseolina* taken from different hosts, but soybeans appear not to be a specialized host. Unlike other fungi, *M. phaseolina* seems to thrive in hot, dry weather conditions, taking advantage of an environment in which plants are already stressed. Thus, it is also known as dry-weather wilt or summer wilt. The fungus can infect soybean plants at any stage of development, but most infections occur within 2–3 weeks of planting (Smith and Wyllie, 1999).

The fungus survives in the soil in the form of small, hardened masses of mycelium known as microsclerotia. Microsclerotia can survive for a few years in dry soil, crop residue, or seeds, but only for a matter of days or weeks in wet conditions. They germinate and infect seeds, seedlings, or plants during periods of warm soil temperature and low soil moisture. Germination occurs on the surface of seeds or roots, allowing germination and infection to continue throughout the season so long as favorable conditions persist. Disease progression is, to a large degree, dependent on weather conditions. If hot, dry conditions persist, infected seedlings may die. During cooler, wetter periods, however, plants may continue to grow and be more or less asymptomatic. Root development and function can be severely reduced by the infection. Above-ground symptoms usually appear in early reproductive stages, especially if the late-season weather is hot and dry. At this stage, the fungus produces black microsclerotia in vascular tissues in the root and stem, producing the characteristic dark grey discoloration that gave the disease its name. The microsclerotia block plant vessels, compromising water and nutrient transport during pod formation and seed filling, resulting in yield loss. The fungus can infect the newly forming seeds, and microsclerotia are released back into the soil from decaying plant residue after harvest, completing the life cycle (Smith and Wyllie, 1999).

Management of charcoal rot emphasizes prevention. Some moderately resistant cultivars are available, but soybean varieties with high levels of resistance are not. Crop rotation can reduce soil populations of microsclerotia, as non-host alternatives are rare. However, this can take multiple seasons, as most alternative crops are merely less susceptible. Measures that reduce crop stress are especially helpful. These include avoiding excessive seeding rates and maintaining good soil fertility and moisture availability. Maintaining plant vigor will reduce the presence of symptoms of charcoal rot (Grau *et al.*, 2004).

Sudden death syndrome

Sudden death syndrome (SDS) was first identified in Arkansas in 1971. It has since spread to all soybean-producing areas in the USA and has also caused significant crop losses in Argentina and Brazil. SDS is caused by the soil-borne fungus *Fusarium solani* f. sp. *glycines*, a specific subvariety of the species *F. solani*. Genetic studies have confirmed that the pathogen is distinct from other forms of *F. solani* (Rupe and Hartman, 1999).

F. solani f. sp. *glycines* mainly inhabits the upper 15 cm or so of the soil and infects soybean plants by entering the roots. Detectable infections have been found as early as 15 days after planting, and infection can continue to occur throughout the growing season. Roots of infected plants are generally stunted and discolored. Stem lesions can appear soon after root symptoms become apparent, but foliar symptoms usually develop in late vegetative or early reproductive growth stages. Leaves show extensive chlorosis and, later, some necrosis; only the major veins remain green. In extreme cases, the leaflets will drop off the plant leaving the petiole attached to the stem. Foliage symptoms are similar to other fungal diseases, so stem symptoms are usually used to make a definitive diagnosis (Grau *et al.*, 2004).

SDS often occurs in combination with SCN. Studies indicate that the presence of SCN is not a necessary part of the disease progression of SDS, but it can make SDS symptoms more severe. Conversely, other investigators have isolated specimens of *F. solani* f. sp. *glycines* from SCN cysts on plants displaying no SDS symptoms. In addition, soil compaction and some aspects of soil fertility are also associated with more severe SDS symptoms (Grau *et al.*, 2004).

The main practice available to farmers for controlling SDS is the use of resistant cultivars. Because of the connection between SDS and SCN, varieties with resistance to both pathogens are valuable in managing SDS. Resistance genes for SDS and SCN appear to be located in the same region of the soybean genome, making the prospect of effective resistance to both diseases promising (Abdelmajid *et al.*, 2007). Delaying planting and planting early-maturing varieties have been shown to reduce the effects of SDS. Improving drainage and relieving soil compaction can also decrease disease risk. Crop rotation has given mixed results, even though SDS shows a fairly high level of host specificity (Rupe and Hartman, 1999; Grau *et al.*, 2004).

Leaf diseases

Soybean rust

Soybean rust is arguably the most destructive disease affecting global soybean production. It is present in all soybean-producing countries and is a more significant problem in tropical and subtropical regions. Rust is especially widespread and destructive in South America and Asia. Rust was first detected in US fields in 2004 and has been responsible for an increasing amount of crop loss since (Wrather *et al.*, 2010).

Two related fungal pathogens, *Phakopsora pachyrhizi* and *Phakopsora meibomiae*, are responsible for soybean rust disease. Of the two, *P. pachyrhizi* is the dominant species and is responsible for most cases. Only part of its life cycle is currently known. Germination and infection occur during cool, wet periods. The most severe rust epidemics result when leaves are wet for 6–12 h/day and temperatures are below 27°C. Infection often happens early in the season, but leaves are susceptible throughout the growing season. The main symptoms are dark lesions on leaves. Fungal reproductive structures grow on the underside of the leaves opposite the lesions. The spores produced are generally spread by wind-blown rain and go on to infect neighboring plants. Lesions may also appear on petioles, stems, or pods, and often produce leaf chlorosis, premature maturity, and defoliation. The impaired leaf function reduces plant vigor and results in yield losses from fewer pods and fewer and lighter seeds (Sinclair and Hartman, 1999c). Yield losses as high as 80% have been reported (Koenning and Wrather, 2010).

The extent of a rust epidemic is strongly influenced by weather and environment. Extended cool, wet conditions are necessary for continued spore production and infection. In the absence of water, spores only remain viable for a matter of a few days. Spores must be spread to neighboring leaves in order to begin reproducing. The alternative host providing refuge after the soybean harvest is not known. Soybean rust is not a seed-borne disease (Sinclair and Hartman, 1999c).

Some resistant cultivars are available. *Phakopsora* spp. have many different races that vary in their virulence. Research is ongoing to identify additional sources of resistance and to transfer these traits to commercial soybean varieties. Fungicide sprays are effective but are economical only when production losses are expected to be significant, i.e. 10–15% or more (Grau *et al.*, 2004).

Brown spot

Brown spot is a common, significant soybean disease worldwide caused by the fungus *Septoria glycines*. *S. glycines* has a moderately wide range of potential hosts, including most *Glycine* spp., several other legumes, and some common weeds. Symptoms mainly involve the leaves, but all plant parts can become infected. Symptoms consist of brown lesions that first appear on cotyledons early in the season and later on the leaves, compromising leaf function and causing premature leaf ageing. During vegetative growth stages, symptoms are usually much less severe, returning and progressing as the plant enters the reproductive stage. As plants approach maturity, the disease can cause premature leaf loss and lesions often appear on stems and pods. Seeds may be infected but usually are asymptomatic. Yield loss is determined largely by the extent of defoliation during pod filling, resulting in fewer and smaller seeds. Symptoms are usually not sufficiently different from those of other diseases to be able to serve as a firm basis for diagnosis (Sinclair and Hartman, 1999a).

S. glycines overwinters in crop residue on the soil surface. Germination occurs during cool, wet conditions in the spring when spores are water-splashed on to emerging plants. If leaves are wet for long periods, the infection can become more severe. Hot, dry conditions during the middle of the growing season are less conducive to growth of the pathogen and may be a factor in making symptoms temporarily less severe. Later in the season, the latent lesions provide infective material for new, more extensive lesions that spread throughout the plant and can then produce spores. Spores and mycelium remain in plant residues after harvest, completing the life cycle (Sinclair and Hartman, 1999a).

Various soybean cultivars differ in their reaction to *S. glycines*, but no completely resistant varieties are known among either domestic or wild species. Minimal tillage, leaving soybean residue on the soil surface, can make the disease more prevalent. Crop rotation is a somewhat effective management tool, but multiple years of non-legume crops are often necessary to significantly reduce the risk of infection. Fungicide sprays during reproductive growth stages can slow disease progression (Grau *et al.*, 2004).

Cercospora leaf blight and purple seed stain

Cercospora leaf blight and purple seed stain are both caused by the fungus *Cercospora kikuchii*. Several other *Cercospora* spp. have been associated with soybean seed discoloration, but *C. kikuchii* is generally regarded as the most significant cause of purple seed stain in soybean fields. Seed stain is usually a more significant loss-producing disease than leaf blight (Schuh, 1999).

C. kikuchii overwinters in infected crop residue, including leaves, stems, and seeds. Only a small percentage of diseased seed, those that are most heavily infected, pass the infection on to the seedling. Severe infection can stunt or kill seedlings, resulting in reduced stands. Most infections result from spores from residue passed to young leaves by wind or rain. Cool, humid conditions and extended periods of leaf wetness facilitate pathogen germination and infection. The infection remains latent until pod filling begins; foliage symptoms usually appear at this time. Small purple leaf lesions tend to grow and coalesce into larger necrotic areas that may include leaf veins. As the disease progresses, the top leaf canopy may defoliate prematurely. However, the infection usually does not spread to the lower leaves, which remain green. The infection does spread to upper stems and pods, and from the pods to the developing seeds. Infected seeds may show staining ranging from small spots to discoloration of the entire seedcoat, or they may be asymptomatic. Heavily stained seeds may experience reduced or delayed germination, but research results are not consistent on this matter (Schuh, 1999).

Premature defoliation is thought to be responsible for some yield loss, but hard evidence for this is sparse. Seed stain is responsible for most economic losses; this is more from reduced grading and value of the crop rather than a direct reduction in yield volume. Nevertheless, relatively moderate but significant yield losses from seed stain occur consistently across the USA and Argentina (Wrather *et al.*, 2001a, 2010; Koenning and Wrather, 2010). The purple seed stain per se does not impair the usefulness of infected seed for processing. However, severe infection can reduce the oil content of the seed, and the stain itself is a grading criterion for commodity soybeans.

The primary management practice for *Cercospora* leaf blight and purple seed stain is to plant less susceptible cultivars. Resistance to the foliar versus seed stain phases of the infection seem to be controlled by separate genetic systems, so development of resistant strains is a more complex task than with other traits. Fungicide application during the pod-filling stage can be effective in slowing disease progression. As with other infections, the presence of surface residue may increase the risk of infection (Grau *et al.*, 2004).

Soybean Production Losses to Pathogens

Losses in the USA

We have a reasonably accurate picture of US soybean production losses to the diseases reviewed above from 1974 to the present. The earliest available data are only for the southern USA, as the initial loss assessment effort was a project of the Southern Soybean Disease Workers (Wrather *et al.*, 1995)[1]. Soybean production in the southern USA constituted a large proportion of total US production, over one-third on average, during the period covered by this initial study, making regional disease losses an important question. As Fig. 2.1 shows, in the late 1980s production in the northern states, the current "soybean belt," began ramping up as southern production held steady, so the relative share of the southern region dropped significantly. Today, these southern states contribute roughly half as much to total US production as they did 40 years ago (USDA, 2018).

In the 1970s, disease was a more significant problem for US soybean production than it is today. In 1974, the first year for which data are available, an estimated 23% of soybean production was lost to diseases. As plant scientists in agricultural research facilities developed disease-resistant soybean cultivars and other disease management practices, losses from diseases steadily declined (Fig. 2.2). In subsequent years, disease losses were consistently below 20% of total production, and by the mid-1990s total losses due to disease had been

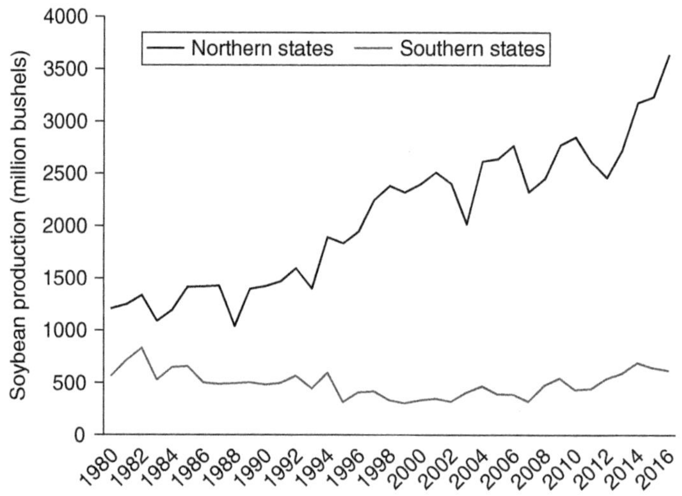

Fig. 2.1. US soybean production from 1980 to 2016. (Data from USDA, 2018.)

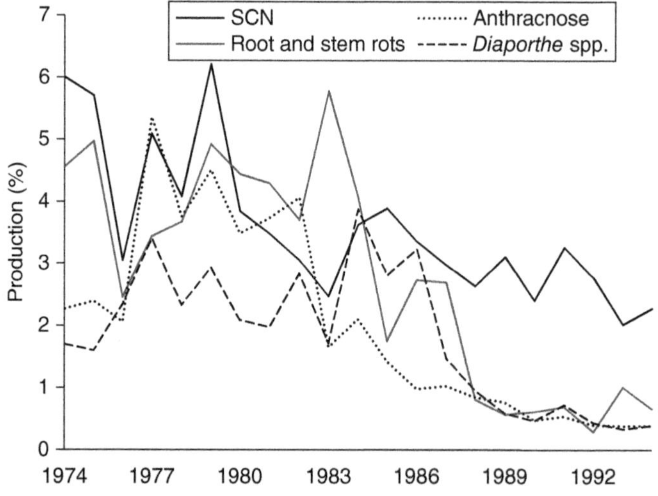

Fig. 2.2. Soybean disease losses in 16 southern US states from 1974 to 1994. (Data from Wrather *et al.*, 1995.)

reduced to less than 10% of potential yield. Throughout this period, SCN was the most serious soybean disease. Up to 1980, SCN caused losses averaging more than 30 million bushels each year, making it responsible for 21% of total losses from disease. As with diseases as a whole, SCN also took a decreasing share of production over time. In 1974, soybean loss to SCN amounted to just over 37 million bushels, or 6% of total production, one of the largest percentage losses on record for this pathogen. SCN losses underwent a gradual decline in succeeding years, not surpassing 4% after 1979, until causing an estimated loss of 14 million bushels in 1994, or 5.4% of production. While SCN was not always the most damaging disease in any given year, it maintained a significant presence in soybean fields throughout the 1980s and 1990s, and has proven to be the most intractable soybean disease to date. As Fig. 2.2 shows, by the early 1990s, SCN was responsible for much larger losses than any other soybean disease in the southern USA (Wrather *et al.*, 1995).

Other diseases have shown much more variation in their impact on soybean producers. Anthracnose was highly prevalent for a few years around 1980 and was the single most damaging disease in two of those years. In 1977, anthracnose accounted for 21% of all disease losses, an estimated total of more than 33 million bushels, or 5% of production. The incidence of anthracnose declined considerably after the mid-1980s; it has been of moderate concern only sporadically since then. Root and stem rots were very serious during the first half of the period, from 1974 to 1984. This class of disease is attributed to a suite of fungi and oomycetes today, notably members of the genera *Phytophthora*, *Pythium*, *Rhizoctonia*, and *Fusarium*, but the specific pathogens involved were not reported in early studies. This suite of pathogens took the top spot in crop loss for a few years in the early 1980s and caused the loss of almost 37 million bushels of soybean production in 1983, or 5.8% of total production. Since then, it has declined to only moderate severity. Blight attributed to *Diaporthe* spp. was intermittently moderate to severe over this period. At its worst, during the period 1977–1986, *Diaporthe* spp. infection caused losses of 2% of production each year, on average. It also declined in later years to a very low level of severity (Wrather *et al.*, 1995).

We have one glimpse into the impact of disease in northern states prior to 1995. Doupnik (1993) estimated overall disease losses for 12 states in the north-central USA in 1989–1991 but did not make separate estimates for individual diseases. For these 3 years, he estimated soybean losses from disease at 12.8%, 16.5%, and 10.1%, respectively, which is roughly comparable to the estimated losses in southern states over the same time frame reported by Wrather *et al.* (1995).

Beginning in 1996, we have a more complete picture of the impact of soybean disease in the USA, and we can compare the relative severity of individual diseases in southern (Fig. 2.3) versus northern (Fig. 2.4) production states over the next 10 years.[2] During this period, SCN continued to be the most consistently serious disease impacting soybean yields in both regions. It was substantially more serious in northern states, taking 8–9% of production in the late 1990s, until it declined in severity to around 4–5% of production through the early 2000s. In southern states, SCN was a more consistent, albeit lesser drain

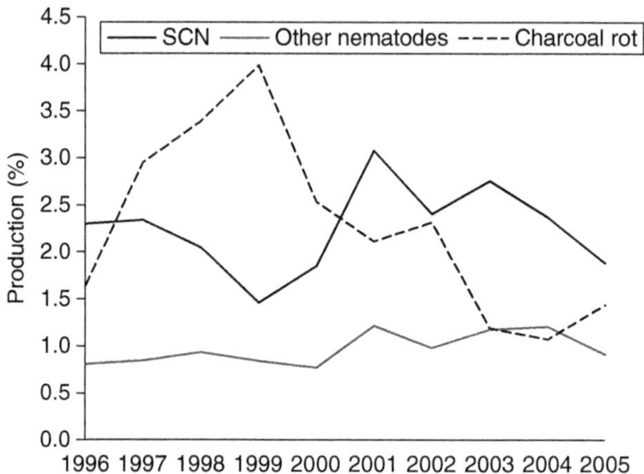

Fig. 2.3. Soybean disease losses in 15 southern US states from 1996 to 2005. (Data from Wrather and Koenning, 2009.)

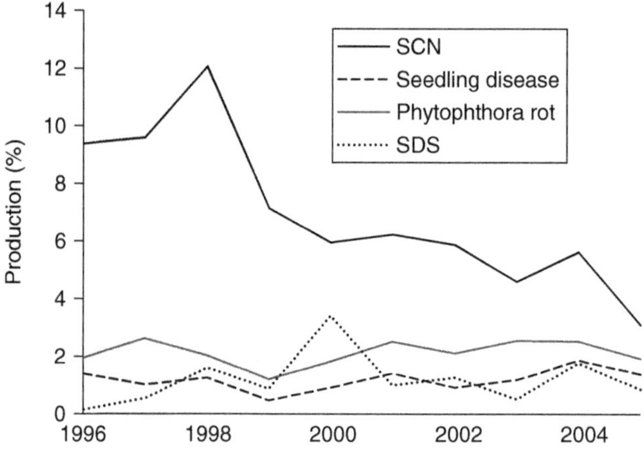

Fig. 2.4. Soybean disease losses in 13 northern US states from 1996 to 2005. (Data from Wrather and Koenning, 2009.)

on production, taking around 2–3% of the crop throughout the period. It was still one of the two most significant sources of crop loss in the south, accounting for 15–20% of disease losses in most years. Other than the primary role played by SCN, northern and southern regions displayed a markedly different pattern of disease incidence and severity. *Phytophthora* root and stem rot was consistently in second place behind SCN in severity in the northern states, although it only occasionally caused the loss of more than 2% of production. Seedling diseases, generally attributed to *Phytophthora*, *Pythium*, and *Rhizoctonia* spp., were also significant, occasionally displacing root and stem rot from its number two position. These two disease classes combine to make *Phytophthora* and other oomycetes the most damaging pathogens preying on soybean plants after SCN in northern states, responsible for losses of 2–3% through this time frame. In Fig. 2.3, we see that the severity of these two diseases tends to rise and fall in concert, as the main difference is growth stage in which the infection becomes most severe. In southern states, on the other hand, oomycetes caused minor losses in production. Here the fungus responsible for charcoal rot, *M. phaseolina*, was the other most consistent threat, its severity surpassing the damage from SCN in some years. In 1999, charcoal rot claimed 4% of soybean production before declining in severity through the early 2000s, when it was responsible for losses of only about 1%. Other types of nematodes, including the root-knot nematode, also consistently took a moderate to severe toll on soybean production, making nematodes overall the most serious class of pathogens in the southern region. In the north, by comparison, root-knot and other nematodes were very minor concerns, while charcoal rot was only occasionally a serious threat. Other fungal diseases were sporadic concerns, responsible for significant losses in one or a few years. For the most part, different fungal pathogens were of primary concern in the two regions. Overall, diseases seemed to claim a slightly smaller proportion of soybean yield in the north than in the south (Wrather *et al.*, 2001b, 2003; Wrather and Koenning, 2006).

After 2005, published reports described losses of soybean production due to disease only for the USA as a whole.[3] Because the southern region now contributes only 15–20% of US soybean production, the aggregate data undoubtedly reflect the disease situation in northern states more accurately than that of the south. In Fig. 2.5, we see that SCN remains the single most damaging pathogen of soybean, responsible for the loss of 3–5% of the soybean crop from 2005 to the present time. *Phytophthora* spp. and other oomycetes, in the form of both seedling disease and rot of more mature plants, are now firmly in second place, consistently taking 2–3% of soybean production. Thus, both classes of pathogens, nematodes and oomycetes, have maintained a high level of severity over the past 20 years and currently account for 45–50% of all disease losses. Charcoal rot shows up as a significant disease over the past few years; it took in excess of 37 million bushels of soybean production annually

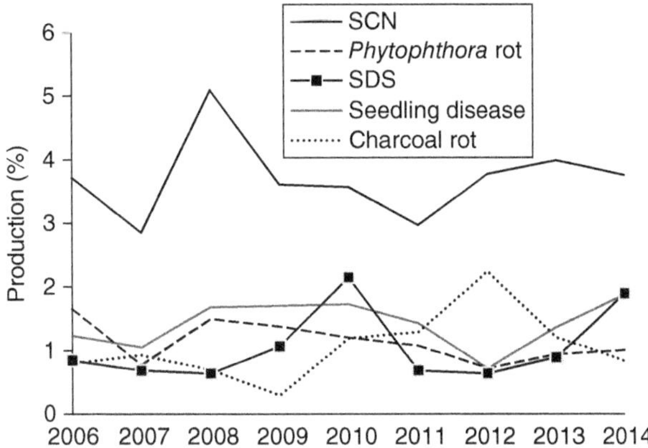

Fig. 2.5. US soybean disease losses from 2006 to 2014. (Data from Koenning and Wrather, 2010; Bradley *et al.*, 2016.)

in 2011–2013. Based on the available data, we cannot say whether this is due to the natural variation in severity in its historic range or if it has spread to new territories.

In the USA, then, we see a pattern over time where a couple of classes of soybean pathogens have become established as the most serious threats to production, while a few others come and go from year to year. After some substantial progress in disease management in the 1980s, overall production losses from disease have fluctuated in a fairly restricted range, at around 10–15%, for the past 20 years.

Global production losses

As soybean agriculture has spread around the world, soybean diseases have followed. There has generally been a time lag, a sort of "honeymoon period," between the introduction of soybeans in a particular region and the appearance of major diseases native to other areas. At the present time, however, soybean is such a well-established crop worldwide that there are few significant differences in which diseases are present in the major producing countries.

Among the more significant diseases, soybean rust is endemic and ubiquitous in Asia. It has been characterized as the most serious soybean pathogen in Asia, causing losses in some countries in excess of 50%. It is not particularly serious in the main soybean-growing area of north-east China but has been known to take serious tolls in south-east China. The New World was spared from this pathogen until it appeared in Brazil and Colombia in 1979 (Oerke, 1999). Nematodes are also significant pests in all major soybean production regions. SCN is present throughout North America and Asia, while root-knot nematodes (*Meloidogyne* spp.) seem to be more prevalent in Latin America. Other significant fungal pathogens, including *Rhizoctonia*, *Fusarium*, *Diaporthe*, and *Septoria* spp., are known in soybean fields around the world with varying degrees of severity in different regions and from year to year. Oomycetes such as *Phytophthora* have been identified in other countries but appear to be a significant threat only in North America. Viruses seem to be more prevalent in Asia than in other regions (Oerke, 1999).

We have only a few snapshots of the quantitative impact of disease on soybean production in countries other than the USA. Each country has a few significant diseases accompanied by a host of others that take smaller tolls on production. In Argentina, brown spot (*Septoria glycines*), purple seed stain (*Cercospora* spp.), and charcoal rot (*Macrophomina phaseolina*) were the most serious diseases in 2006, the most recent year for which data is available, taking 2.6%, 2.4%, and 2% of soybean production respectively (Table 2.1). SDS completed the group of most significant disease threats, although at 1.7% of production its severity could be characterized as moderate (Wrather *et al.*, 2010). In previous years, overall losses from disease were less and the profile of severe diseases was somewhat different. SDS was among the more serious diseases in both 1994 and 1998 but was responsible for losses of only 1% or less of production. Nematodes were one of the most severe diseases in 1994 and

Table 2.1. Soybean yield reduction in Argentina as a result of disease (in million bushels). (Data from Wrather *et al.*, 1997, 2001a, 2010.)

Disease	1994	1998	2006
Anthracnose	1.35	1.35	1.66
Bacterial diseases	0.45	0.68	0.83
Brown spot	1.34	2.03	43.18
Brown stem rot	1.35	2.70	0.83
Charcoal rot	0.60	0.68	33.21
Diaporthe–Phomopsis complex	3.14	0.00	0.00
Downy mildew	0.45	0.00	0.00
Frogeye leaf spot	0.00	0.00	0.83
Phomopsis seed decay	0.00	2.70	1.66
Phytophthora rot	0.60	3.38	2.49
Pod and stem blight	0.00	1.35	6.64
Purple seed stain	0.90	1.32	39.86
Root-knot and other nematodes	5.83	2.70	0.83
Sclerotinia stem rot	6.73	15.53	4.98
Seed disease	0.00	0.68	1.66
Seedling disease	0.90	1.32	1.66
Southern blight	0.60	0.00	0.00
Soybean cyst nematode	0.00	2.03	0.83
Soybean rust	0.00	0.00	1.66
Stem canker	0.45	4.73	0.00
Sudden death syndrome	4.93	5.40	28.23
Viral diseases	3.14	2.03	1.66
Other diseases	0.45	0.68	1.66
Total disease losses	33.18	51.26	174.38
Total soybean production	455.58	716.43	1487.97
Loss to disease (%)	6.80	6.68	10.50

accounted for a production loss of 1.2%. The only disease that caused more than a 2% yield reduction in 1994 or 1998 was *Sclerotinia* stem rot in 1998, which was also in the top spot in 1994 (Wrather *et al.*, 1997, 2001a). Overall, the available data indicate that disease management was not a significant producer concern in Argentina until more recent years. It would appear that only as the soybean production area has expanded into more disease-prone locations has yield loss increased to a level closer to the world average. However, the available data are insufficient to determine conclusively whether this is a long-term trend or just a snapshot of normal year-to-year variation.

In recent years, soybean rust (*P. pachyrhizi*) has become the most severe disease in Brazil (Table 2.2). It was responsible for a loss of 7.3% of production in 2006, accounting for well over half of all disease losses. This stands in stark contrast to the 1990s, when records show no significant production loss due to rust. Losses from other pathogens have been kept more in check. Purple seed stain caused yield losses of 2.5–3% in the 1990s, but only a 1.1% loss in 2006. No other disease took more than 1% of production in 2006. Brown spot, which took 5.7% of yield in 1998, charcoal rot, responsible for a 2–3% loss in the

Table 2.2. Soybean yield reduction in Brazil as a result of disease (in million bushels). (Data from Wrather *et al.*, 1997, 2001a, 2010.)

Diseases	1994	1998	2006
Anthracnose	2.84	0.07	8.07
Bacterial diseases	0.00	0.00	4.77
Brown spot	40.37	80.55	12.48
Brown stem rot	0.23	0.00	1.47
Charcoal rot	34.13	27.53	13.21
Diaporthe–Phomopsis complex	2.28	0.00	0.00
Downy mildew	0.00	0.00	1.84
Frogeye leaf spot	0.00	0.00	1.47
Fusarium root rot	0.00	0.00	1.47
Phomopsis seed decay	0.00	0.09	4.77
Phytophthora rot	0.29	0.00	0.00
Pod and stem blight	1.14	23.01	0.00
Powdery mildew	0.00	5.75	0.00
Purple seed stain	34.13	34.52	26.42
Rhizoctonia aerial blight	0.15	0.53	11.01
Rhizoctonia–Pythium root rot	1.14	0.00	0.00
Root-knot and other nematodes	2.84	11.51	9.54
Sclerotinia stem rot	5.51	0.06	7.34
Sclerotium blight	0.00	0.00	2.13
Seed disease	5.69	0.07	2.57
Seedling disease	1.84	0.06	3.01
Soybean cyst nematode	11.38	17.62	19.08
Soybean rust	0.00	0.00	173.22
Stem canker	66.06	0.37	0.00
Sudden death syndrome	0.55	7.34	11.74
Viruses	0.00	0.00	3.67
Other diseases	0.73	4.77	0.00
Total disease losses	211.29	213.85	319.29
Total soybean production	906.49	1192.75	2091.90
Loss to disease (%)	18.90	15.20	13.24

1990s, and stem canker, which caused a loss of 5.9% of production in 1994, have all been reduced to low levels of severity. Nematodes, especially SCN, seem to be on the rise in Brazil and infection levels may be severe in some locations, although all nematodes in aggregate caused only a 1.2% production loss nationwide in 2006. Overall, total losses to disease stood at 18.9% in 1994, 15.2% in 1998, and 13.2% in 2006 (Wrather *et al.*, 1997, 2001a, 2010).

Reliable data for losses from disease in China are only available for 1994 and 1998. In these two years, rust caused yield losses of 1.4% and 1.1%, respectively, a very moderate toll. SCN, in contrast, was consistently responsible for the largest disease losses, taking roughly 4% of production annually in those years. Among other fungal diseases, downy mildew, frogeye leaf spot, and *Fusarium* rot claimed 1–3% of soybean production in the 1990s. Finally, in contrast to the situation in the other major producing countries, viral diseases were responsible for significant production losses in China, taking 2.6% of production in 1994

Table 2.3. Soybean yield reduction in China as a result of disease (in million bushels). (Data from Wrather *et al.*, 1997, 2001a, 2010.)

Disease	1994	1998
Anthracnose	0.00	0.61
Bacterial diseases	0.00	0.61
Brown spot	0.00	0.00
Brown stem rot	0.00	0.00
Downy mildew	13.35	6.69
Frogeye leaf spot	18.53	18.25
Fusarium root rot	8.10	9.12
Phomopsis seed decay	0.00	0.00
Phytophthora rot	0.00	0.30
Pod and stem blight	0.00	3.04
Cercospora purple seed stain	0.00	1.82
Rhizoctonia blight	0.00	0.61
Root-knot and other nematodes	0.00	0.61
Sclerotinia stem rot	2.84	3.04
Seed disease	0.00	3.04
Seedling disease	0.00	3.04
Soybean cyst nematode	25.96	27.37
Soybean rust	9.30	7.30
Viruses	16.78	21.29
Other diseases	0.00	6.08
Total disease losses	94.86	112.84
Total soybean production	561.88	540.52
Loss to disease (%)	14.44	17.27

and 3.3% in 1998 (Wrather *et al.*, 1997, 2001a, 2010). For some reason, viral diseases seem to be present at a much higher degree of severity in most Asian countries compared with the Americas or Europe (Oerke, 1999).

Oerke (2006) also estimated that pathogens of all types decreased world-wide production of soybeans by 18.7% below its potential in 2001–2003[4]. This estimate is roughly comparable to that calculated in the studies by Wrather. Overall, diseases seem to take a fairly consistent 10–20% share of production in the main producing countries. This is true despite each growing region having its own characteristic profile of common diseases, and a disease that is serious in one country may be trivial or largely absent in another. Each regional profile can be somewhat stable over time, although individual diseases come and go and their relative incidence and severity can change from year to year. As the following section explains, a variety of environmental factors can play a determining role in the disease experience of a particular field in any given year.

Loss Variability and the Conditioning Effects of the Environment

One reason behind the observed variability of production losses across pathogens and geographies and over time is the conditioning effects of the environment. It is a fundamental fact of plant pathology that a favorable environment is a necessary condition for the progression of disease. Even if a pathogen and a plant host come into intimate contact, in the absence of appropriate environmental conditions, largely defined in terms of temperature and moisture, infection and disease cannot happen. Moisture and temperature interact to strongly influence the production of infectious agents in fungi (Sun and Yang, 2000). Moreover, if either of these conditions is outside the acceptable range for a given pathogen species, germination and infection cannot occur, even if the other is at a level that would otherwise be optimal. Within these bounds, temperature and moisture have independent, as well as interactive, effects.

Many pathogens are most vigorous at temperatures that are different from the optimum temperature for development of the host plant. This is highly adaptive for the pathogen, allowing it to attack when the host is in a relatively weaker state. Many fungi germinate and grow best at temperatures below those that soybeans find most conducive to growth. Thus, the infection can take hold earlier in the year when soybeans are at a younger, more vulnerable stage of development. In other cases, the most rapid disease progression takes place at a temperature that is optimal for neither the pathogen nor the host but is one where the pathogen holds a relative advantage.

Moisture is necessary for fungal spore germination and also plays an important role in the spread of disease. Some fungi, such as rust and brown spot, require more or less extended periods of continuous leaf wetness in order to successfully infect a soybean plant. In cases such as these, temperature and moisture interact. As a rule, longer periods of leaf wetness allow germination to take place at a wider range of temperatures; as temperatures depart from the optimum, germination takes longer, but only up to a point. For most fungi, there are maximum and minimum temperatures beyond which germination will not occur, regardless of moisture levels. In a parallel fashion, in the absence of adequate moisture, germination is also prevented, regardless of temperature. Many soil-borne pathogens, such as *Phytophthora* spp. and nematodes, also require the extended presence of liquid water in order to travel through the soil and infect the roots of young plants. In contrast, other fungi, the powdery mildews for example, are inhibited by excess moisture and thrive in hot, dry conditions as long as there is some dew at night (Agrios, 2005).

Other environmental conditions can affect disease progression. Wind can play an important role in the spread of disease. For example, wind patterns from a tropical storm system helped spread southern corn leaf blight from its origins in Iowa to several mid-western and southern states in 1970 (Rosenzweig *et al.*, 2001), and Hurricane Ivan is credited with carrying soybean rust from South America to Louisiana in 2004 (Luck *et al.*, 2011).

Soil pH affects some diseases but has no effect on others. It can directly impact the virulence of some pathogens and can also influence host plant health and nutrition. A paucity or abundance of any of a long list of soil nutrients can make plants more or less susceptible to various specific diseases. Herbicides and air pollutants can have indeterminate effects on disease by aiding or inhibiting pathogen growth, moderating plant susceptibility, or modifying the general soil environment (Agrios, 2005).

Summary

Estimating the proportion of agricultural production lost to pests is a challenge for many reasons, not least of which is the counterfactual condition: what production would have been in the absence of pest loss is not observable and must be estimated. Nevertheless, techniques for doing so exist and can provide reasonably accurate estimates of relative pest impacts. Indeed, the total impact is considerable: an estimated 25–40% of agricultural production is lost to pests each year worldwide, despite control efforts. Estimating the production lost to diseases is inherently more challenging due to their diversity. Pathogens display significant differences in basic biology and physiology, mode of attack, preferred environmental conditions, and other characteristics. It is appropriate in many contexts to speak of weeds or insects in the aggregate; it is much less so for pathogens. There is much variation in disease incidence, severity, and the resulting yield loss. This variability plays out across seasons and geographies. Overall, it is estimated that 10–20% of global soybean production is lost to disease each year.

Notes

[1] Wrather *et al.* (1995) estimated soybean yield losses due to disease in Alabama, Arkansas, Delaware, Florida, Georgia, Kentucky, Louisiana, Maryland, Missouri, Mississippi, North Carolina, Oklahoma, South Carolina, Tennessee, Texas, and Virginia.

[2] In Wrather *et al.* (2001b) and later reports, Missouri moved from the southern to northern category. The northern category contains Illinois, Indiana, Iowa, Kansas, Michigan, Minnesota, Missouri, Nebraska, North Dakota, Ohio, Pennsylvania, South Dakota, and Wisconsin. The southern category contains Alabama, Arkansas, Delaware, Florida, Georgia, Kentucky, Louisiana, Maryland, Mississippi, North Carolina, Oklahoma, South Carolina, Tennessee, Texas, and Virginia.

[3] Koenning and Wrather (2010) reported losses from 27 states. All states included in earlier reports are included in this paper, except that no estimated losses in Mississippi are present.

[4] As noted earlier, Oerke (2006) groups nematodes with animal pests rather than pathogens. He reports an 8.9% production loss to pathogens and 8.8% to animal pests. Since insects cause only very small losses in soybeans, nematodes account for the bulk of animal pest losses in these estimates. Here, we added these two estimate categories, making this loss estimate more comparable to those in the studies by Wrather, although it may slightly overestimate total disease losses.

3 Disease Incidence, Severity, and Conditioning Factors

To limit the potential yield losses from disease, farmers must anticipate the presence of pathogens in their fields and use appropriate practices to control them. As we describe throughout this book, these activities are not straightforward. In order to understand how soybean farmers form expectations about the incidence and severity of disease, how they choose among different control practices, and how they decide when to use them, we focus our empirical analysis on a specific class of soybean diseases – seedling disease and mid-season root rots – and a specific region – the USA. Like all soybean diseases, their incidence, severity, and impact on yield are variable and difficult to anticipate. Strategies for their control are broadly applicable to other soybean diseases. As such, insights from this more focused analysis can be generalized to other classes of diseases and other regions. We begin this chapter by examining the biological details of the selected diseases (the underlying pathogens, their lifecycles, etc.), the factors that condition their incidence and severity in any given year and agricultural field, and the disease control strategies that might be available to soybean farmers.

Disease Description

Seedling disease, or damping off, is the result of infection by one of a few different pathogens. The pathogen may infect seeds prior to germination or seedlings before the first trifoliate leaf stage (V1). Infected seeds appear soft and rotten and do not germinate. Infected seedlings may or may not emerge. The infected hypocotyl may appear swollen and water soaked, and becomes too weak to support the seedling after emergence. The cotyledons may also be affected. Roots are generally severely stunted. The infection completely disrupts the physiological processes of the seedling and results in its death. In severe cases, fields must be replanted (Schumann and D'Arcy, 2010).

If infection occurs later in the season, plant death is less likely but still possible. Infection enters through the roots, as with seedling disease, and can spread throughout the plant. Symptoms can appear at almost any time from early vegetative to late reproductive growth stages (Grau *et al.*, 2004). The earliest symptoms are usually stunting and discoloration of roots, which are difficult to observe. Above-ground growth is stunted as the pathogen robs the plant of nutrients. Stem lesions and leaf chlorosis can also appear. In severe cases, premature leaf death and defoliation can cause plant death (Rupe and Hartman, 1999; Schmitthenner, 1999).

A fungal pathogen commonly responsible for damping off is *Rhizoctonia solani*. *R. solani* can be saprophytic (able to consume dead vegetable matter), so it can overwinter in crop residue or, if no host plants are available, remain dormant in the soil as small, hard masses of mycelium known as sclerotia. It can be spread by moving water or by transfer of contaminated soil. *R. solani* often infects soybeans early in the season, as the optimum temperature for germination and infection for most strains is 15–18°C (59–64°F). The pathogen thrives, and causes the most severe infection, under moderate soil moisture conditions. Soil that is dry or waterlogged inhibits its growth. In less severe cases, soybean plants can continue to grow but exhibit chlorosis and stunting later in the season (Yang, 1999b). The species consists of several distinct strains known as anastomosis groups (AGs). The groups are differentiated based on the type of interaction that occurs when individuals come into contact with one another in the soil or in a host plant. AG-4 is the most common strain infecting soybean fields, and is also virulent to maize, which limits the effectiveness of crop rotation in controlling *R. solani* (Yang, 1999b; Agrios, 2005). Not all strains of *R. solani* are pathogenic, however. In one survey, Murillo-Williams and Pedersen (2008) found improved root health and yield associated with *R. solani* infection. Apparently, a non-pathogenic strain of this fungus was out-competing virulent strains of other soybean seedling pathogens.

Several members of the fungal genus *Fusarium* have also been connected with seedling disease. *Fusarium* spp. tend to favor somewhat warmer soils and can infect soybean plants later in the season; in particular, *Fusarium oxysporum* causes blight and *F. solani* is the pathogen behind SDS, a very serious mid-season disease (Nelson, 1999a; Rupe and Hartman, 1999). Both of these are also known to infect seeds and seedlings, causing seed and root rot and damping off (Nelson, 1999b).

More recently, *Fusarium graminearum* has emerged as a significant pathogen of soybean seedlings. *F. graminearum* is also an important pathogen of maize and wheat seedlings and, like its relatives, causes significant losses later in the season due to ear, head, and stalk infections in maize and wheat. Its recent rise to prominence seems to be largely a result of the increasing popularity of conservation tillage and earlier planting dates. While both of these practices offer agronomic benefits, they entail seeds being planted into cooler, wetter soil. This delays germination and increases the opportunity for infection (Broders *et al.*, 2007b). The cool, wet conditions seem only to increase the exposure of seeds and seedlings, however; studies indicate that temperature has only a minimal impact on the virulence of *F. graminearum* (Ellis *et al.*, 2010). Its wide range of hosts means that the common maize/soybean and maize/soybean/wheat rotations are not effective in controlling this pathogen. Other *Fusarium* spp. have been isolated from diseased soybean seedlings, but the results of pathogenicity tests suggest that these may be opportunistic secondary colonizations rather than actual pathogenic infections (Rizvi and Yang, 1996).

The most common causal agents of soybean seedling disease and mid-season rot are not fungi, but oomycetes. These resemble fungi, but certain details of their biology make them more closely related to brown algae and kelp. However, oomycetes do not perform photosynthesis as algae do. Fungal cell walls contain chitin, the same protein that makes up the hard outer shells of insects and crustaceans, but no cellulose. Nearly all oomycetes, in contrast, have cell walls made of cellulose and related compounds, with no chitin (Agrios, 2005). In addition, the filament cells of oomycetes are diploid, having a complete double set of chromosomes, like plants and animals, while fungal cells are haploid, with only a single set of genetic material. Oomycetes also share some important metabolic traits with kelp and diatoms, which further differentiate them from fungi (UCMP, 2006). One crucial result of these differences is that oomycetes are unaffected by most common fungicides and require specialized chemicals for control. Because of their similarities to fungi in overall structure and disease progression, however, oomycetes are categorized with fungi in plant pathology and the chemicals used for their control are commonly called fungicides.

Probably the most common oomycete pathogens responsible for damping off are members of the genus *Pythium*. With well over 100 identified species that are virulent to a multitude of crop and non-crop plants, this is generally referred to as the *Pythium* complex. Individual *Pythium* spp. are usually virulent to many different host plants. The genus is found worldwide (Agrios, 2005). *Pythium ultimum* is perhaps the most frequent pathogen of soybeans (e.g. Schlub and Lockwood, 1981), but in most surveys multiple species are isolated from infested soils or infected plants. In a wide-ranging census covering 11 states responsible for 77% of US soybean production, Rojas *et al.* (2017a) isolated 84 oomycete species from symptomatic soybean seedlings, of which 69 were various representatives of the *Pythium* complex. Overall, 95% of all oomycetes isolated in that census were members of some *Pythium* sp. However, not all species were found in all locations.

Pythium diseases are monocyclic, with the pathogen completing only one life cycle generation per year. *Pythium* spp. can live saprophytically, and thus can overwinter in crop residue, or can survive as dormant oospores in the soil for many years. Like most water molds, *Pythium* spp. require water-saturated soil for germination and infection, as the zoospores can only move through free water. The species most often responsible for soybean seedling disease in the USA germinate at temperatures of 10–15°C (50–59°F). Other species germinate at much higher temperatures and are more prevalent in warmer climates (Yang, 1999a), while the growth rate of still other species appears to be almost unaffected by temperature. There is also evidence of temperature influencing fungicide sensitivity in several species (Matthiesen *et al.*, 2016). One phylogenic study found that the *Pythium* spp. that flourish in warmer soil are closely related to *Phytophthora* spp., which also prefers warmer conditions (Levesque and de Cock, 2004). As noted earlier, the cool, wet conditions that are most favorable to most *Pythium* spp. also place soybean seeds and seedlings at greater vulnerability of infection and tend to occur more often in the early spring. Thus, early planting and conservation tillage seem to promote seedling disease from *Pythium* spp. in addition to other pathogens. Flooding around the time of emergence has been shown to significantly reduce soybean stand populations and increase *Pythium* spp. populations, while flooding later in the season did not have those effects (Kirkpatrick *et al.*, 2006).

Unlike their *Phytophthora* cousins, *Pythium* spp. generally have a wide range of potential hosts. One study found 14 species virulent to maize as well as soybeans (Dorrance *et al.*, 2004). Other studies have confirmed this general condition, although there is often substantial variation; any given species is likely not equally pathogenic to both crops (Broders *et al.*, 2007a). This condition does not seem to be due to genetic variation within individual *Pythium* spp., with different races or strains pathogenic on different hosts. Instead, individual isolates appear to be capable of infecting multiple hosts (Zhang and Yang, 2000). This characteristic implies further challenges in disease control, as the standard maize/soybean crop rotation may do little to limit *Pythium* spp. populations.

Perhaps the most economically damaging oomycete belongs to the genus *Phytophthora*. While many field surveys recover much greater numbers of isolates from the *Pythium* complex, there is some uncertainty as to whether this represents actual differential infection rates or is an artifact of sampling procedures. In any event, *Phytophthora* spp. tend to be among the most aggressive of soybean pathogens (Rojas *et al.*, 2017a). Although some of the more than 80 species of *Phytophthora* have fairly broad host ranges, most are host specific.

Infection of soybean plants by *Phytophthora sojae* was first documented in Indiana in 1948, although the cause of the disease was unknown at that time. Similar symptoms were recorded in other mid-western states over the next few years, and the pathogen was first isolated from diseased seedlings in North Carolina and Ohio in 1954 (Schmitthenner, 1985). *P. sojae* was first described and named by Kaufmann and Gerdemann (1958). Soybean varieties resistant to *P. sojae* were first identified in 1954, and the *Rps* gene isolated from them was bred into popular cultivars over the next 10 years. The resistant cultivars were planted throughout the US soybean production area, and by the late 1960s, the pathogen appeared to have been defeated. In the early 1970s, however, the genetic flexibility of *P. sojae* became apparent. During this decade, a number of newly virulent *P. sojae* races were discovered in mid-western fields. By 1993, 27 distinct races had been identified (Schmitthenner *et al.*, 1994; Yang *et al.*, 1996). Currently, at least 14 *Rps* genes at eight genetic loci have been identified in soybean. At the same time, at least 12 distinct avirulence genes have been identified in *P. sojae*, giving rise to some 55 races currently known to plant pathologists (Malvick and Grunden, 2004; Tyler, 2007).

Like all oomycetes, *P. sojae* is heterotrophic; it must consume other organic tissue in order to survive. It is primarily a parasitic pathogen, infecting soybean roots and stems. It can also survive as a saprophyte, consuming dead crop residue. The primary means of dispersal and infection are aquatic zoospores, produced asexually. Zoospores can swim through flooded soil and are attracted to chemicals released by soybean roots. When a zoospore contacts a root, it changes into an adhesive cyst and germinates to produce hyphae filaments that penetrate the root wall and begin the infection. *P. sojae* can also reproduce sexually; the fertilized egg, called an oogonium, develops into an oospore. Oospores may germinate about 30 days after they are formed, but can also survive dormant in the soil for many years, awaiting the proper conditions before germinating and producing hyphae (Grau *et al.*, 2004). This requires both an adequate amount of soil moisture and a sufficiently high temperature. Minimal oospore germination has been observed at a soil temperature of 15°C (59°F), but a soil temperature of approximately 25°C (77°F) is optimal for germination and growth (Schmitthenner, 1999; Tyler, 2007).

P. sojae can attack soybean plants at any stage of development. Infected seeds may rot without germinating. Young seedlings may turn brown, wilt, and die before emerging. Damping off may occur after emergence. In this case, the young roots turn brown and the lesion extends up the hypocotyl until it collapses and the seedling dies. Pre- and postemergence damping off can be misdiagnosed as water damage, as the infection often occurs in fields flooded just after planting. Plants that are infected when they are older may display stem lesions and leaf chlorosis to varying degrees, and may or may not die, depending on the degree of tolerance or resistance of the particular cultivar. In all cases, the roots will show stunting, discoloration, and rot. Root damage may lead to opportunistic infections by other fungi. In highly tolerant cultivars, root rot may be the only symptom, possibly accompanied by some stunting or slight chlorosis of the plant. *P. sojae* infection can still reduce the yield of such plants by as much as 40%. The wide range of observable symptoms greatly complicates diagnosis and thus treatment decisions (Schmitthenner, 1985, 1999).

Incidence and Severity: Spatial and Temporal Variability

While both mid-season root rot and seedling disease and their causative pathogens are known throughout the US soybean production region, there is much variation in both the presence of the pathogens and the severity of resulting disease. This variation exists on many levels. For example, several disease surveys have discovered a non-uniform distribution of fields from which oomycete pathogens have been recovered. Figure 3.1 illustrates that fields with *P. sojae* infestations can occur in clusters in some areas, while other areas nearby can remain unpopulated. The open circles on the map of North Dakota represent fields from which samples were taken, while solid circles denote those fields where *P. sojae* isolates were obtained (Nelson *et al.*, 2008). This variability of incidence may be a reflection of many factors that can promote or inhibit pathogen populations, such as soil texture and structure (Ghorbani *et al.*, 2008).

Pathogen population profiles can also vary geographically; Rizvi and Yang (1996) found that oomycetes made up 42–100% of isolates from soybean fields

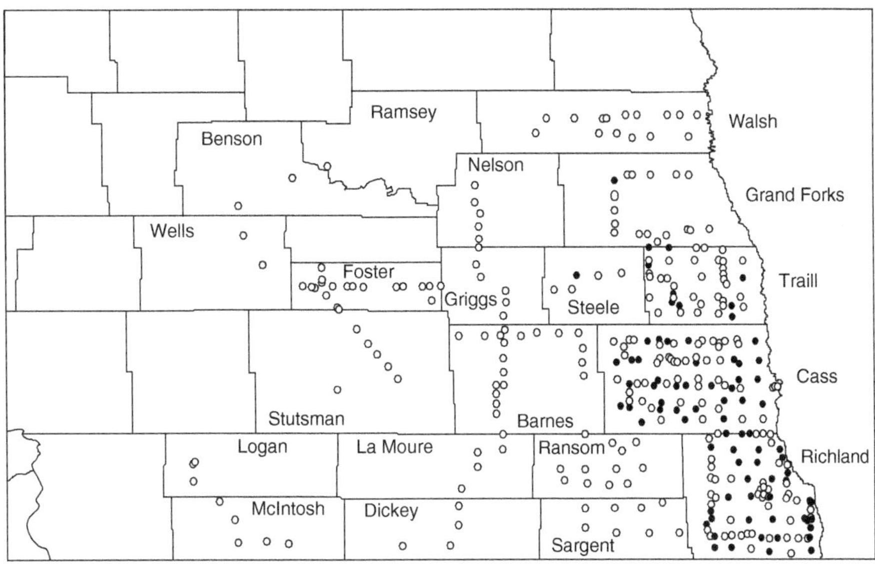

Fig. 3.1. Incidence of *Phytophthora* spp. in North Dakota. (From Nelson *et al.*, 2008.)

in Iowa, with *Fusarium* and *Rhizoctonia* spp. constituting the remainder of fungal-type pathogens recovered. In a more extensive survey conducted by Rojas *et al.* (2017a,b), seedling disease pathogens of almost 3500 samples were characterized. For each sample, they determined the incidence of each oomycete species and the disease severity (i.e. relative pathogenicity) under various environmental conditions, such as temperature. They found a wide diversity of populations and expected virulence in the samples, across both time and geography. Figure 3.2 shows this variation for the primary pathogenic species across the Corn Belt. It also helps to visualize how neighboring geographies tend to have more similar community profiles than more widely separated locations. This is likely attributable to the finding that temperature, precipitation, soil clay content, and soil water content were the main factors influencing community composition (Rojas *et al.*, 2017b).

If we look at disease severity, defined as the estimated percentage of yield lost to disease, at a national scale, we see some regularities.[5] Oomycete diseases seem to be more serious in northern states. As Fig. 3.3 shows, Ohio (7.8%), Michigan (6.5%), Wisconsin (4.7%), Minnesota (3.4%), North Dakota (6.6%) and South Dakota (3.4%) experienced the greatest yield losses from seedling disease and mid-season root rot during the period 2011–2015. However, Kansas (4.0%) also had significant losses in a very different climate, and Iowa and Nebraska (0.6% each) had much smaller losses than their equally northern neighbors. If we disaggregate the data into the two separate disease processes, early season seedling disease and mid-season root rot, the picture becomes somewhat clearer. As seen in Fig. 3.4, *Phytophthora* root rot (PRR) was significantly more serious in northern states, and more uniformly so as well. Michigan (2.1%), Wisconsin (3.3%), Minnesota (1.5%), North Dakota (1.7%), and South Dakota (2.0%) had the greatest percentage losses, while some southern states

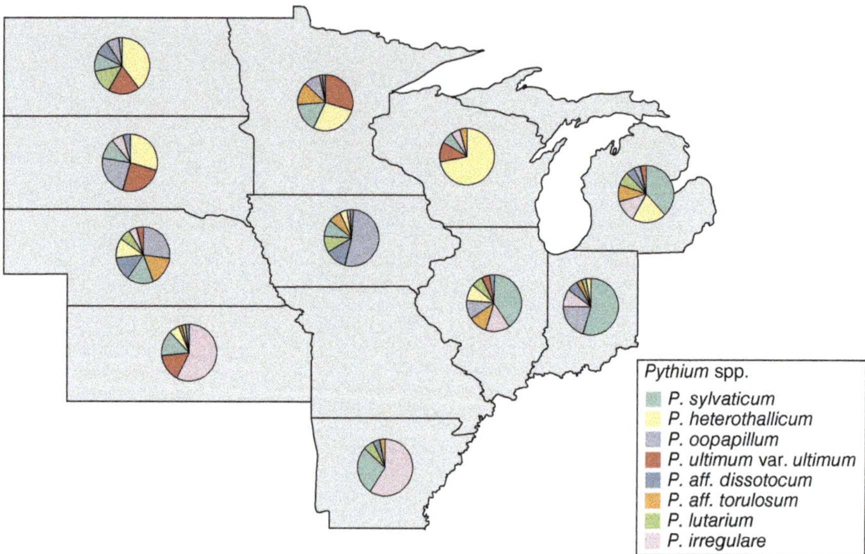

Fig. 3.2. Distribution and abundance of the top eight pathogenic oomycete species across the states sampled in 2011 and 2012. (From Rojas *et al.*, 2017b.)

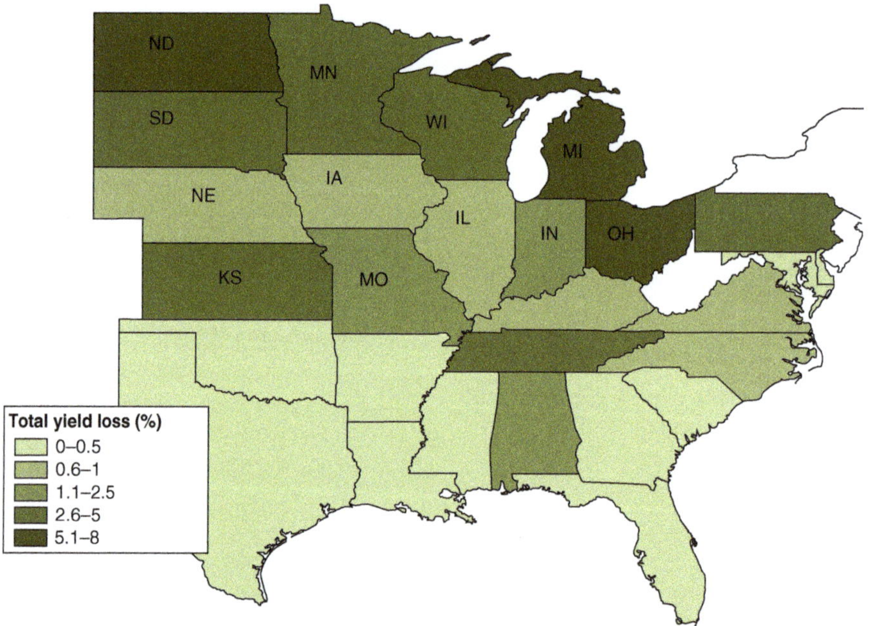

Fig. 3.3. Yield loss to oomycete diseases. (Data from Bradley *et al.*, 2016.)

had none. The severity of seedling disease, however, is geographically mixed. Figure 3.5 shows the estimated percentage yield losses from seedling disease. North Dakota (5.0%), Michigan (4.5%), and Ohio (4.0%) had significant losses from seedling disease as well as root rot, while Kansas (3.8%) and Pennsylvania

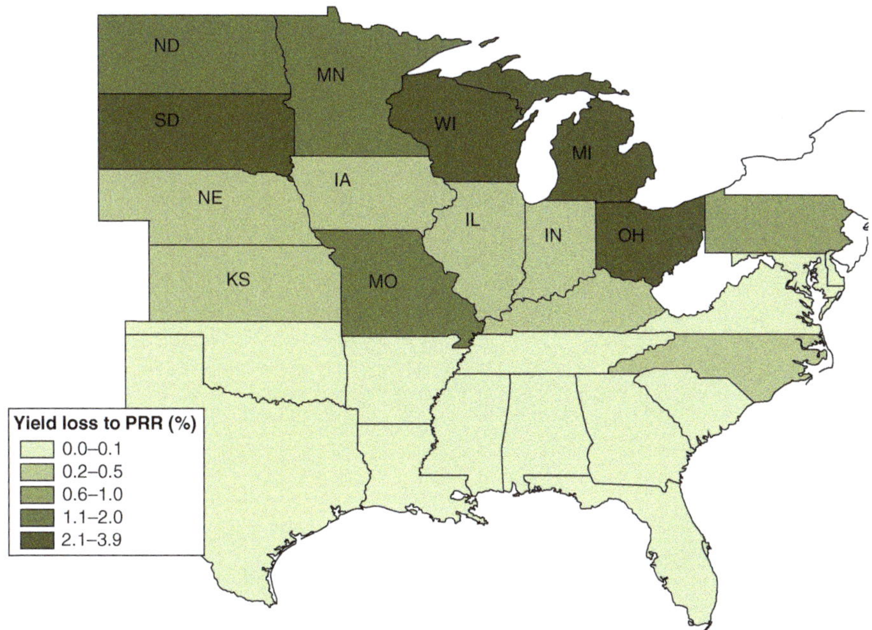

Fig. 3.4. Yield loss to PRR. (Data from Bradley *et al.*, 2016.)

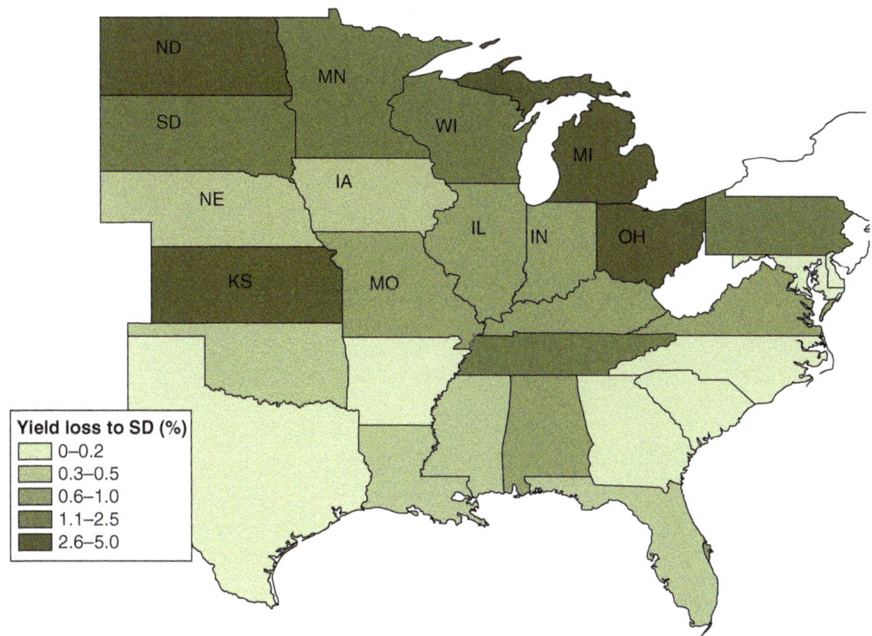

Fig. 3.5. Yield loss to seedling disease (SD). (Data from Bradley *et al.*, 2016.)

(2.0%) were in the top five for seedling disease severity but experienced very little loss from root rot.

In addition to geographical variation, individual states also experience different levels of disease losses from year to year. Figures 3.6 and 3.7 show estimated losses for the five states with the most severe levels of PPR and seedling disease, respectively. Disease severity can change greatly from one year to the next. For example, root rot in Wisconsin was responsible for only an estimated 1.9% yield loss in 2011 but claimed 8.7% of soybean produce in 2012. In Pennsylvania, seedling disease was not a factor in soybean loss in 2013 but

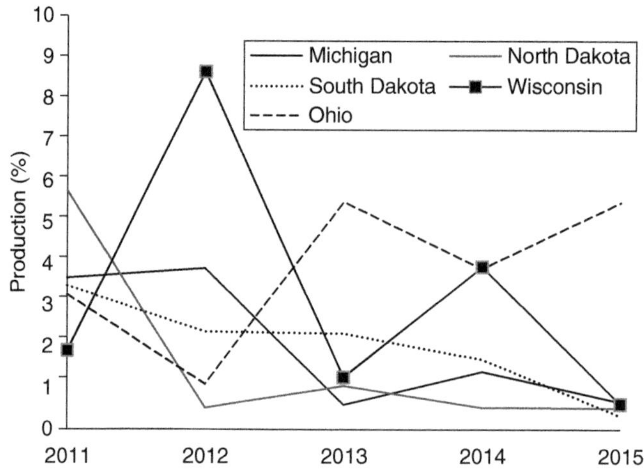

Fig. 3.6. Estimated yield losses to PRR from 2011–2015. (Data from Bradley *et al.*, 2016.)

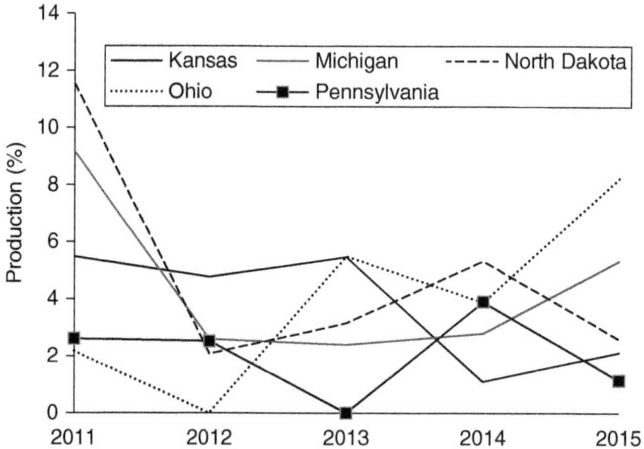

Fig. 3.7. Estimated yield losses to seedling disease from 2011–2015. (Data from Bradley *et al.*, 2016.)

took 3.9% of the crop in 2014. The spatial and temporal variation outlined here has been documented on many levels, from individual fields to regions of the world (e.g. Ryley *et al.*, 1998; Grijalba and Gally, 2015). This serves to emphasize the importance of each farmer's localized understanding of the incidence and severity of such disease in informing the overall management decision process.

Factors Influencing Incidence and Severity

Environmental factors

Characteristics of the overall environment can influence the incidence and severity of seedling disease. The weather and basic soil type and structure are the chief variables in this category. As a rule, however, these are not under a large degree of farmer control.

Weather affects seedling disease in two ways: rainfall and temperature. Heavy spring rains can waterlog the soil, creating prime conditions for oomycete germination and infection. Improving soil drainage can help to limit the problems caused by waterlogged soils. Small adjustments in planting date can also have an impact by avoiding either waterlogged soils or temperatures conducive to pathogenicity. Some research indicates that moderate rain or irrigation that occurs soon after planting can promote plant growth and maturation more than pathogen growth, enhancing disease resistance (Dorrance *et al.*, 2009). The exact timing and amount of water can be critical, however. Kirkpatrick *et al.* (2006) found that a 3-day flood at the time of seedling emergence significantly reduced soybean stands, at least partly due to seedling disease, while a 7-day flood at growth stage V4, when the young plants have four trifoliate leaves on the main stem and are usually 20–25 cm tall, had no effect. Likewise, the choice of planting date can work to discoordinate seedling and pathogen development to the benefit of the crop. A soil temperature of 25°C (77°F) is optimal for *Phytophthora sojae* germination (Schmitthenner, 1999), while some *Pythium* spp. can be virulent at soil temperatures as low as 13°C (55°F) (Rojas *et al.*, 2017a). This is the main reason why *Pythium* spp. are primarily responsible for early season seedling disease, while *P. sojae* is more common as a cause of mid-season root rot. The further along developmentally the soybean crop is by the time the soil reaches a temperature where early season pathogens can be virulent, the less susceptible it will be to seedling disease.

The type of soil present in a particular field can strongly influence how challenging disease control might be there. Soil with greater water-holding capacity can promote germination, and coarser-grained soil enables greater zoospore mobility. Soil that is poorly aerated due to compaction has been associated with increased oomycete disease. Soil management techniques that reduce compaction and increase drainage can reduce disease incidence and severity by encouraging root formation and penetration, and thereby plant health (Ghorbani *et al.*, 2008).

The specific soil type found in a given agricultural field plays a determining role in how the field responds to precipitation or warming spring temperatures. Soil scientists classify soils into 12 major types based on their parent rock, age and degree of weathering, mineral content, and other criteria. As Fig. 3.8 shows, there is considerable soil type variation across North America, sometimes over fairly short distances. This typology, however, is only a convenient way of communicating general information about individual soils. Soil composition in the field exists along a continuum, not in discrete types. Thus, there are also a number of levels of subcategories for soil classification, necessary to accurately describe an individual soil sample. Among the second tier of classification criteria are the average soil moisture regimes. Figure 3.9 shows the variation in moisture regimes across North America. These regimes interact with the basic soil types to produce considerable regional variation in soil moisture, temperature, and warming rate. The heterogeneity, however, extends even down to the level of the individual field, with significant variation in soil characteristics within as well as among agricultural fields. This degree of local heterogeneity makes each farmer's local knowledge of field conditions a critical body of information in formulating disease control strategies and choosing among them (Arnold, 2005).

Cultural and agronomic practices

The characteristics of the agricultural environment have as much of an impact as those of the natural environment on the incidence and severity of seedling

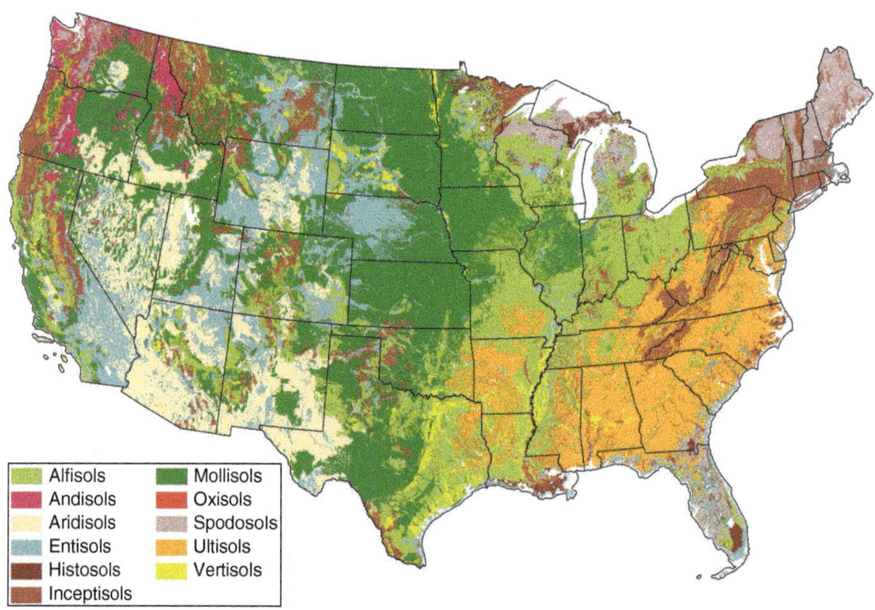

Fig. 3.8. Major soil types in the USA. (From USDA Natural Resources Conservation Service, www.nrcs.usda.gov/wps/portal/nrcs)

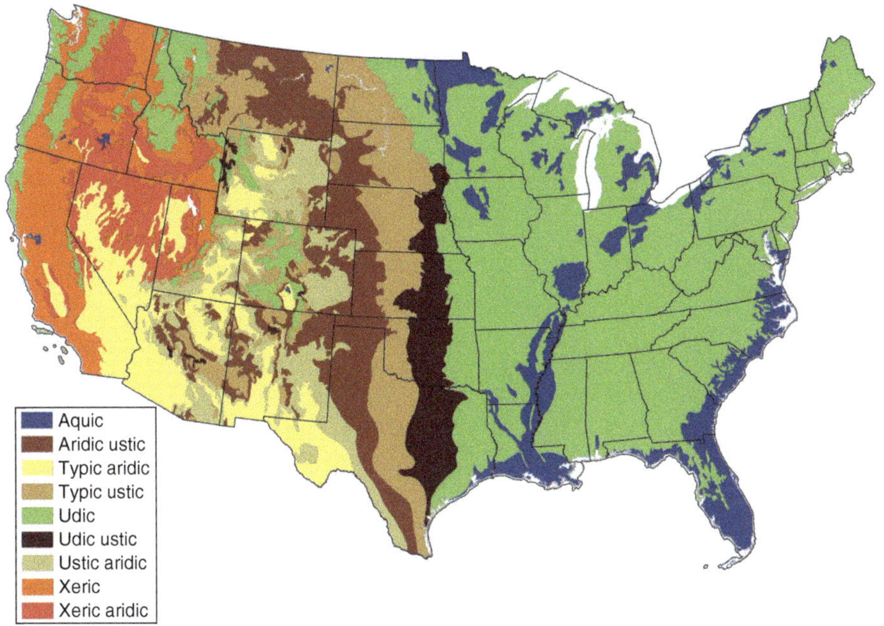

Fig. 3.9. Soil moisture regimes in the USA. (From USDA Natural Resources Conservation Service, www.nrcs.usda.gov/wps/portal/nrcs)

disease. The practice of agriculture essentially creates an artificial environment in production fields with the aim of maximizing the growth and production of crop plants while minimizing the growth of all organisms that might interfere or compete with crop growth. The specific methods and practices used to do this, and the specific characteristics of the environment that is created, can have a significant effect on the prospects for disease incidence and severity in any given production season. Specifically:

- Soil drainage is critical in controlling oomycete infestation (Schmitthenner, 1985). Oomycetes require saturated soil for germination and zoospore mobility. Installing drain tiles can decrease the chances of seedling disease. When irrigating, it is recommended that only sufficient water to maintain crop growth be applied (Ghorbani *et al.*, 2008).
- Tillage of soil can improve drainage, as well as directly inhibit oomycetes by breaking up hyphae and burying oospores (Rush *et al.*, 1997; Zhang and Xue, 2010). However, gains in disease control are by no means certain; while tillage may decrease seedling disease, it might make the soil environment more hospitable for SCN (Workneh *et al.*, 1999).
- Crop rotation can offer some benefit in controlling *P. sojae*, as soybeans are the only host crop (Zhang and Xue, 2010). A change in predominant rotation practice, moving to more sequential years of soybeans, has been cited as a factor in the increased disease incidence of that pathogen (Dorrance *et al.*, 2009). Crop rotation can also improve general soil quality and thereby crop health, but its utility for seedling disease suppression may be minimal

as oospores can survive for many years dormant in the soil, awaiting suitable conditions for germination (Grau *et al.*, 2004).

- Fungicide seed treatments have been shown to be effective in suppressing seedling disease (Bradley, 2008). Higher seeding rates and closer row spacing seem to decrease root disease but at the cost of potentially promoting foliar disease (Rush *et al.*, 1997).
- Choice of a planting date is important and is driven by many considerations, one of which is the goal of having the crop pass through the seedling stage while soil temperature and moisture are not conducive to pathogen growth (Chaube and Singh, 1991).
- Use of fertilizer can improve crop nutrition, reducing stress and enhancing disease resistance (Chaube and Singh, 1991). However, overfertilization can leave excess nutrients for pathogens, as well as promoting prolific growth and delaying crop maturation, which can exacerbate disease severity (Ghorbani *et al.*, 2008).
- Promoting general soil quality in terms of ample organic matter, microbial diversity, and proper soil pH may also help to decrease disease severity (Ghorbani *et al.*, 2008).
- Appropriate sanitation measures can slow the spread of disease. In particular, cleaning equipment to ensure that infected soil or crop debris is not transferred from one field to another has been shown to be effective (Gould and Hillman, 1997).

Pest–host interactions and dynamics

Seedling disease and mid-season rots are the result of an interaction between pathogens and individual soybean plants, mediated by the natural and agricultural environment, as described above. As such, the genetic profiles of the pest and host populations, how they interact, and the dynamics of how they change over time are critical factors in determining the incidence and severity of seedling disease in a given soybean field. Soybean genetics are largely intentionally developed and refined by laboratory scientists and breeders. The choice of a specific cultivar is an important disease management decision point for any farmer. The genetics of the pathogen populations, in contrast, are shaped by natural processes. Novel traits arise from mutation and recombination, and their survival is in the hands of natural selection. The producer's choice of cultivar is at once an adaptation to those genetics and a source of selection pressure on the pathogen population.

Consider the pest–host interactions of cultivated soybean cultivars with *P. sojae* for example. Developing resistance to *P. sojae* in the domestic soybean, *Glycine max*, has been a challenge for soybean breeders, with only a little help from Mother Nature. Soybeans originated in China (Qiu and Chang, 2010), but *P. sojae* appears to be a North American native, using members of the genus *Lupinus* as hosts prior to the introduction of soybean agriculture (Förster *et al.*, 1994), so the two species are unlikely to have encountered each other before that time. Nevertheless, plant scientists have found several varieties of soybean with varying levels and types of resistance, including some wild and domestic strains from Asia (Hymowitz, 1984). As described earlier, some genes, known as *R* (resistance) genes, confer complete immunity to particular *P. sojae* races. To date, at least 14 *Rps* (resistant to *P. sojae*) genes have been mapped at eight separate locations in the *G. max* genome, and a few of these genes have been incorporated into commercially used soybean cultivars. In turn, *P. sojae* races virulent to common *R* genes have increased over time, so stacks of multiple *R* genes are becoming a more important disease control tool (Dorrance *et al.*, 2003a). However, the structure of the soybean genome limits the potential for stacked resistance traits. For example, only one of six possible alleles can be present at locus 1 in any given cultivar; the other five must remain unused. This adds to the importance of the other weapon in the genetic arsenal, partial resistance. Partial resistance does not endow immunity but lessens the severity of the infection. However, it is effective against all *P. sojae* races. Thus, a high level of partial resistance is a valuable complement to race-specific resistance in genetic methods of soybean disease control.

Changes in soybean population genetics come about only slowly. Oomycete disease resistance is not the only trait producers and breeders are interested in. For commodity soybean production, harvestable yield is an overriding concern; seed traits are valuable only insofar as they contribute to yield or decrease production costs. When breeders add a new trait, such as an additional *R* gene, they must ensure that existing desirable traits, such as herbicide tolerance, emergence, maturity, growth characteristics, resistance to other diseases, are also transferred to the new germline and not compromised by the new trait. They must also ensure that no undesirable traits are transferred into the new germline from the source of the *R* gene.

Producing a new soybean variety with a stable set of optimal traits can require multiple backcrosses and extensive field testing. The difficulty of this process is compounded by the nature of soybean biology. Since the soybean is a self-pollinating plant, all crosses must be performed manually, which makes breeding all the more tedious, time consuming, and labor intensive. It takes 6–10 years, on average, to develop a new soybean variety to the point where it is ready for commercialization after the sources of desirable genetic traits have been identified and isolated (Stine Seed Company, 2016).

In contrast to its *G. max* host, *P. sojae* is highly variable genetically, and populations can adapt rather quickly to changes in their environment, in particular to newly resistant hosts. As described earlier, *P. sojae* is an obligate parasite; it must infect a living host in order to complete its life cycle and produce a new generation. As such, the criteria for biological success are quite simple: survival is based primarily on virulence. Wide genetic diversity within a population is thus highly adaptive, as there is more likely to be at least some individuals in the population that can infect whatever host might be present. In this context, diversity is manifested as the presence of avirulence (*Avr*) genes that interact on a gene-for-gene basis with race-specific resistance genes in soybeans (*R* genes) and allow an individual zoospore to infect a soybean plant with a particular *R* gene. The presence of a specific *Avr* gene or set of *Avr* genes defines a race of *P. sojae*.

A large degree of genetic diversity is not merely a theoretical possibility for *P. sojae*. Survey studies, repeated over many years, in Ohio (Schmitthenner *et al.*, 1994; Dorrance *et al.*, 2003a), Illinois (Malvick and Grunden, 2004), Iowa (Yang *et al.*, 1996; Murillo-Williams and Pedersen, 2008), Michigan (Kaitany *et al.*, 2001), and North Dakota (Nelson *et al.*, 2008), as well as regional studies (Workneh *et al.*, 1999; Dorrance *et al.*, 2016) and some results from other countries (Drenth *et al.*, 1996; Ryley *et al.*, 1998; Grijalba and Gally, 2015) all show similar trends:

- The number of distinct races detected in plant and soil samples has increased over time, indicating increasing diversity.
- Later studies tend to identify races virulent to *R* genes that have not been deployed in the area from which the samples were collected.
- Later studies tend to identify individual isolates that do not fall into previously described virulence races, and the proportion of isolates in this category seems to be increasing over time.
- Later studies identify isolates that are virulent to more *R* genes. Some studies have found isolates virulent to all known *R* genes.

These results suggest that the presence of multiple *Avr* genes does not come at any significant survival cost to *P. sojae*, so that the degree of diversity and complexity is only likely to continue to increase. This also suggests that a disease control strategy based solely on race-specific resistance may not achieve long-term success. Instead, most experts recommend an integrated disease control program utilizing soybean varieties with specific resistance against the most problematic races present in a particular region in concert with partial resistance, judicious use of chemical control measures, and appropriate agronomic practices designed to minimize the incidence and severity of seedling disease and root rots. These results also demonstrate the importance of regular field surveys to determine the genetic profile of pathogen populations, so that farmers may be better able to choose among disease control practices (Schmitthenner, 1985; Dorrance *et al.*, 2003a).

Pest–soil microbe interactions and dynamics

The battle between *P. sojae* and *G. max* is not the only evolutionary struggle going on in agricultural fields, and not the only one that impacts disease incidence and severity. Soils harbor a large and diverse microbial population with an intricate web of coevolutionary interaction. Pathogens are a part of this population, and the interactions also extend to crops. There is much that is unknown about soil microbes. Studies indicate that there may be as many as 10^9 microbe cells in each gram of soil; less than 5% of the species represented in that population have been cultured and studied (Garbeva *et al.*, 2004). Research in this area is ongoing and is strongly driven by the observation that some soils seem to be naturally suppressive to crop diseases. Thus, in some soils, the requisite factors of a pathogen, a susceptible host, and the proper abiotic environmental conditions such as moisture and temperature are all present but disease does not develop. Disease suppression, or the lack thereof, appears to be the result of coevolutionary interaction among pathogens and other soil microbes, i.e. the selection pressure for a particular trait does not come solely from one-to-one competition between two species but from competitive interaction with multiple species. This opens up the possibility that disease incidence and severity is influenced by the totality of the soil microbe population and that this population, as well as the suppression that might result from a beneficial microbe population profile, may be amenable to soil management practices (Kinkel *et al.*, 2011).

Robust coevolutionary processes are a key part of soil health, an age-old concept that has only recently attracted efforts to formulate a rigorous definition and develop a research agenda. Soil health is generally viewed as a subset of ecosystem health and is defined in terms of resilience to stress. Higher levels of disease suppression and biodiversity are crucial components of that resilience (van Bruggen and Semenov, 2000). Janvier *et al.* (2007) defined health as the capability of the soil to function as a living system, sustain biological productivity, and promote plant, animal, and human health. Two facets of disease suppression are widely recognized; both are the result of healthy soil and the biodiversity it promotes. General suppression is linked to the total amount of microbial activity and competition. In a diverse, competitive community, there are no unfilled niches and no one species, such as a pathogen, can come to dominate the population. With population numbers thus kept in check, the opportunities for infecting crops are more limited. Specific suppression is the result of the presence of specific organisms that are antagonistic to particular pathogens, usually by means of releasing toxic or antibiotic compounds (Janvier *et al.*, 2007; Chaparro *et al.*, 2012). Biodiversity also implies the existence of a diverse pool of many small microbe populations, in addition to the main collection of species responsible for most biological activity in a soil community. This reserve pool is available to respond to stress in such a manner that, because of its diversity, again precludes any one species from coming to dominate the community (van Bruggen and Semenov, 2000; Garbeva *et al.*, 2004).

The rigorous study of soil health is in its infancy, however, and has so far progressed little beyond demonstrating that there are solid connections among biodiversity, biological productivity, and disease suppression, and that healthy soil is an exceedingly complex system. Attempts to identify specific microbe species or soil traits that are key sources of disease suppression or to apportion causal responsibility among the various factors described above have been largely unsuccessful. It seems clear enough, however, that soil health is a factor in crop disease management that could contribute to disease control (van Bruggen and Semenov, 2000; Garbeva *et al.*, 2004; Janvier *et al.*, 2007).

Summary

The incidence and severity of soybean disease in any given field and year results from a constellation of many diverse factors. Some of these – agronomic practices and the cultivar planted – are under the farmer's control, but choices are made based on broader agronomic and economic criteria than just disease control. Environmental factors are consequential, are not generally controlled, and are difficult to anticipate with any degree of accuracy. Pest interactions with hosts and various soil microbes add yet another level of uncertainty. Thus, as a practical matter, disease incidence and severity are highly stochastic. Yet farmers must form expectations about them in order to decide which disease control practices to implement, if any, every year. We examine the farmer's decision process in the next chapter.

Notes

[5] Compilation of the data set used here was initiated by the Southern Soybean Disease Workers in 1974 and was continued for many years by Dr. J. Allen Wrather, University of Missouri, and Dr. Steve Koenning, North Carolina State University. Data has been published by Wrather *et al.* (1995), Wrather and Koenning (2009) and Koenning and Wrather (2010). National data since 1996 is available at http://extension.cropsci.illinois.edu/fieldcrops/diseases/yield_reductions.php. The dataset is currently managed by Dr. Tom Bradley, University of Illinois, Dr. Tom Allen, Mississippi State University, and Dr. Paul Esker, Earth University, Costa Rica, who generously supplied it for our work.

4 An Economic Framework of Disease Management

The standard economic model of pest management is the damage abatement model, as initially developed by Lichtenberg and Zilberman (1986). At the core of the damage abatement model is the concept of the economic threshold (ET) as a decision criteria. The idea that there is a specific, non-zero pest population that triggers the implementation of control measures was first discussed by Stern *et al.* (1959). They defined the economic injury level (EIL) as the lowest pest population density that would cause economic damage, defined as an amount of damage that would justify the cost of control. Given the likely time lag between identification of the population density and application of pest management measures, they also defined the ET as the population density at which to make the decision to control, in order to prevent the population from reaching the EIL. The ET is necessarily a lower population density than the EIL.

The above definition was conceptually satisfying but suffered from one flaw. As Davis and Tisdell (2002) pointed out, it was a break-even concept, based on the difference between the total expected loss from a specific population density and the total cost of pest management measures that would keep the population at or below that density. Headley (1972) proposed a refinement to the original model, which was further developed by Hall and Norgaard (1973), bringing it more into line with standard economic theory. He defined the EIL as the population density where the monetary value of the marginal loss from a given pest is just equal to the marginal cost of control. The difference between the two concepts is subtle but important, as the latter is the form that leads to profit maximization. Given that pest management measures are necessary, the EIL represents the optimum pest population density. Maintaining the population at either a greater or lesser density will result in a smaller net return after management expenses.

Headley (1972) also developed and expanded the ET concept in a number of ways. He formally observed that at very low pest population densities,

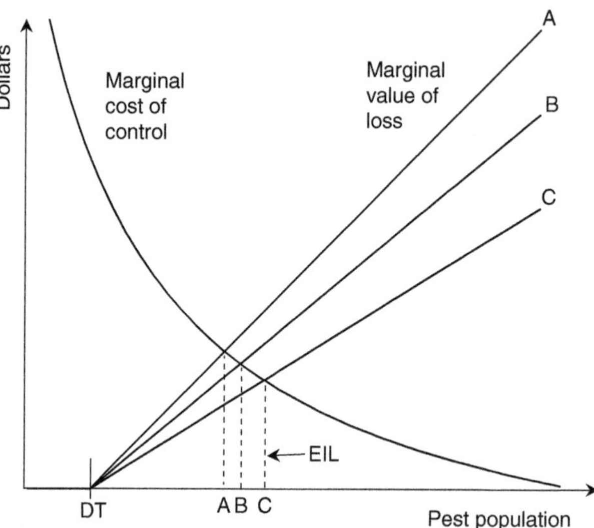

Fig. 4.1. Relevant thresholds in soybean disease management. A, higher crop value; B, medium crop value; C, lower crop value. (Adapted from Headley, 1972.)

expected damage is zero; plants that are otherwise healthy can sustain a certain level of infection or infestation before yield is affected. At some point, however, the pest population will produce a measurable loss of yield. This density level has come to be known as the damage threshold (DT) (Weersink *et al.*, 1991; Nutter *et al.*, 1993). The actual population level corresponding to the DT will be different for different pests and crops. For instance, as described in Chapter 2 (this volume), soybean plants can withstand a considerable amount of physical injury from insect pests before experiencing yield loss. Nematodes, in contrast, can cause measurable yield loss at low populations, even before above-ground symptoms are apparent. In most cases, the EIL will be greater than the DT.

Finally, Headley (1972) explicitly showed that the magnitude of the EIL is dependent primarily on the relative prices of production output and pest management inputs. Figure 4.1 illustrates this, as well as the relative levels of the DT and EIL. As the pest population density increases from zero, when it reaches the DT the damage becomes noticeable and the economic value of the yield loss is measurable. As successive incremental increases in the pest population take successively larger portions of the crop, the value of the loss mounts. For a higher-value crop, (Fig. 4.1, line A), the cost of the loss increases more quickly and meets the cost of control at a lower level of physical loss than for medium- (line B) or lower-value (line C) crops. If the cost of control increases, the marginal cost of control curve would shift to the right. Thus, pest management measures would become economical only at higher levels of physical damage and loss value, which would be depicted by shifting the EIL to the right.

This model illustrates that, except for rare and unusual cases, pest eradication is not economically justifiable. Even if the marginal cost per dose remains constant, the increasing quantities of control measures needed to achieve successive decreases in the pest population mean that the marginal cost of control, as a function of the population, increases dramatically with lower densities. Thus, in general, the cost of eradication of any particular pest is excessive with currently available pest management technologies.

The Damage Abatement Model

The ET concept was a useful concept for discussing pest management decision making, but it was not initially integrated into the overall farm management decision process. As this was recognized, early modeling attempts entered pest management inputs like any other agronomic inputs into the standard agricultural production function. Typical use of the standard Cobb–Douglas production function resulted in overvaluation of pest management inputs and prescriptions of excessive use (Sexton *et al.*, 2007). Specifically, standard production models were of the basic form:

$$Y = g(Z) \tag{4.1}$$

where total production, Y, is defined by a yield function $g(Z)$, in which Z is a vector of agronomic inputs (e.g. fertilizer, irrigation, and seeds) and pest control inputs (e.g. pesticides).

As Lichtenberg and Zilberman (1986) pointed out, however, pest management does not increase potential yield directly and in some cases may in fact decrease it slightly; resistant cultivars may not have yield characteristics comparable to other varieties, and chemical pesticides may cause slight amounts of crop damage. Instead, pest management inputs prevent damage, defined as yield loss, from pests and thus should enter decision models separately. In this context, the level of damage is directly related to the pest population present in a particular field; the relationship between these two levels is known as the damage function. If we define damage, D, as the proportion of yield lost to pests, we can express it as a function of the pest population at a specific point in time, N, or $D(N)$.

Entering pest management inputs as a separate damage abatement term requires strict limits on how pest management can influence production. Abatement cannot be less than zero, in which case the crop suffers the maximum amount of damage possible, given certain pest population and environmental conditions at a particular place and time. This would be the outcome when no pest control measures were applied. Neither can abatement be greater than one, the case where pest damage is completely prevented and full potential yield is realized. Application of pest control measures in excess of the level required to attain complete abatement will, at best, have zero marginal benefit. In some cases, the marginal benefit may be negative, such as when a pesticide is also toxic to the crop at high levels or in the case of environmental degradation due to runoff (Waterfield and Zilberman, 2012).

Modeling pest control inputs through a damage abatement term had two main advantages (Sexton *et al.*, 2007). First, while it is well established that all factors of production have decreasing marginal productivity, damage abatement factors do so in a significantly different manner. Disease management, for example, does not directly repair or mitigate crop damage; rather, it decreases the population of pathogens that are responsible for the damage. Thus, the marginal factor productivity of pest control inputs decreases more quickly than in the case of conventional factors of production. Standard models of factor productivity thus systematically overestimated the marginal benefit of damage control inputs and led to recommendations of overuse. Modeling pest management inputs as damage abatement corrected this error by maintaining a more appropriate valuation. Second, the productivity of pest management can also decrease over time, as resistance develops and spreads in pest populations. In conventional production functions, lower marginal factor productivity decreases demand for, and use of, that factor. For damage abatement inputs, however, decreasing marginal productivity can temporarily increase demand as farmers attempt to maintain an acceptable level of control. This may continue until some alternative factor becomes more cost-effective and replaces the less effective pest control input. The damage abatement specification allows this contrary effect of declining effectiveness.

The total quantity of crop produced then becomes the product of the potential yield, the amount attained in the case of no pest damage, and the proportion of that yield realized due to pest management efforts, i.e. the level of successful damage abatement. Thus, the standard damage abatement model expressed total production as:

$$Y = g(Z)\left[1 - D(N)\right] \tag{4.2}$$

In this model, Z is a vector of agronomic inputs unrelated to damage control, such as labor, fertilizer, fuel, and so forth. The relevant pest population causing damage, N, is itself a function of the initial population, N_0, the quantity of pesticide applied, X, and alternative control measures, A (e.g. genetic resistance in the seed, cultural controls). These factors come together in the kill function, h, which is specified as:

$$N = h(N_0, X, A) \tag{4.3}$$

The kill function gives the pest population density after control measures have been implemented. It is decreasing in X and A.

The farmer's profit, Π, is thus the product of the price received multiplied by the total quantity produced less production costs. The farmer's decision task is to maximize profit, or:

$$\max_{Z,X,A} \Pi = pg(Z)\left[1 - D\left(h(N_0, X, A)\right)\right] - uZ - wX - vA \tag{4.4}$$

where u is the cost of agronomic inputs, w is the unit cost of the pesticide application, and v is the cost of alternative pest control methods. For a given set of agronomic inputs, Z, farm profit will be maximized at a point where the value of the last increment of pest control measures applied is just equal to the value of the increment of damage prevented thereby; the pest population density, N, that results in this relationship is the EIL. The precise mix of pest management methods used can change based on relative prices. The model predicts that farmers will use the pest management practice that has the lowest cost, not per volume unit but per unit of damage prevented. Thus, the relative effectiveness of alternative pest control practices also enters into the decision process.

Empirical implementation of the damage abatement framework in agricultural pest control analysis has often been hindered by a lack of theoretical guidance on the appropriate functional form to use for the damage or kill functions. Carrasco-Tauber and Moffitt (1992) and Praneetvatakul *et al.* (2002) each compared logistic, Weibull, and exponential specifications of the damage function with a standard Cobb–Douglas specification of marginal productivity in pest control. Both studies confirmed that the Cobb–Douglas model overestimates marginal productivity, and that logistic and Weibull damage functions produced similar results. In both cases, the exponential specification produced a lower, and seemingly more realistic, estimate of the marginal productivity of pesticides, but authors have continued to note that there is still no firm theoretical basis for preferring one specification over another.

Uncertainty and the Damage Abatement Model

The standard damage abatement model described above treats all variables of interest as deterministic. This is a convenient simplification, but relaxing this assumption permits useful extensions. Allowing for stochasticity in the pest population, for instance, gives rise to considerations of preventive versus responsive pest control practices. Responsive methods are based on knowledge of actual pest populations that occur in each crop cycle, while preventive methods are implemented before that knowledge is available and thus are based on expectations of the future population.

Preventive pest management

When farmers use preventive pest management measures, they must make a judgement as to the probable future pest population, based on history and their knowledge of relevant conditions on their land. They then must choose a level of pest control input, X (e.g. fungicide seed treatment), that will maximize their expected profit from preventive management, Π_p, given the particular pest population and its expected damage. Following Sexton *et al.* (2007), the profit-maximization problem of the farmer can be specified as follows:

$$\max{}_X \Pi_p = pg(Z)\int_{N_1}^{N_2}\left[1-D\left(\bar{N},X\right)\right]\Psi\left(\bar{N}\right)dN - wX \tag{4.5}$$

Here, $\Psi(N)$ is the pest population density function, which ranges from N_1, the smallest population that can sustain itself, to N_2, the carrying capacity of the relevant area. $\bar{N}$ denotes the expected pest population. Note that X is chosen for a given $\bar{N}$ and does not change based on what the actual pest population turns out to be.

Responsive pest management

In a responsive system the profit from management is not based on a chosen X, but instead the function $X(N)$, which is determined based on the actual observed pest population. The farmer also incurs a monitoring cost, m, in order to determine the value of N. Maximizing profit from responsive pest management, Π_r, may then be represented by:

$$\max_{X(N)} = pg(Z)\int_{N_1}^{N_2}\left[1-D\big(N,X(N)\big)\right]\Psi(N)dN - w\int_{N_1}^{N_2}X(N)\Psi(N)dN - m \quad (4.6)$$

The farmer will choose between the two management modes based on the relative profit from each and the availability of relevant information. Whether Π_p or Π_r offers the greater profit to the farmer comes down to the tradeoff between monitoring cost and the pest control input savings obtained by monitoring. If monitoring is relatively less expensive, the farmer will have an incentive to implement a responsive program (Sexton *et al.*, 2007). It is worth emphasizing here that the implication of Eqns 4.5 and 4.6 is that if pesticides are relatively inexpensive, farmers may use preventive measures even when the year-to-year variation in $X(N)$ is large, making accurate prediction difficult. Put a little differently, if the information needed to make an accurate assessment of the pest population is difficult or impossible to obtain in a timely manner, this is the equivalent of a high monitoring cost and likewise pushes farmers toward a preventive pest management program (Stern, 1973).

In disease management practice, the choice of preventive or responsive management is often based on the type of disease present. Foliar diseases, such as rust, generally appear later in the season and can be treated by responsive means, such as spraying with fungicide after the infection becomes apparent. In contrast, many soil-borne diseases, including seedling disease and mid-season rots, must be prevented, mostly through the use of genetically resistant varieties and other preventive methods. In these cases, either the disease progresses too quickly for the farmer to mount an effective response or observable symptoms manifest too late to be used as decision criteria. Some diseases can be controlled by employing both preventive and responsive methods in some fields and seasons, depending on the particular disease conditions that prevail.

Farmers' risk preference

In addition to temporal variation in the pest population, other variables relevant to the pest management decision are also stochastic. Pannell (1990) described several of these. The pest population may vary across and within fields, and in areas of higher density, pests must compete more with each other for resources. Thus, with the same total population size, the damage per pest, and thus total loss, may be smaller from populations with more variable densities. The effectiveness of pesticides and other control practices can be uncertain as well, due to environmental conditions at the time of application, the characteristics of the pest population, or an inherent capacity to control the pests. These and other factors can have individual, interactive, and sometimes conflicting effects on the final calculation of the ET and management decisions.

In the presence of uncertainty, farmers' risk preferences become important. The role of risk has been examined in the context of pest management decisions. Feder (1979) pointed out that pesticides and other pest control practices perform an insurance-like function, decreasing the variation in the range of possible outcomes, even if the mean expected outcome does not change. In contrast to other types of inputs in other industries, pest management measures are thus risk reducing; a risk-averse farmer will respond to the presence of risk with an increased quantity of pest control input use. Moffitt (1986) noted, however, that if a risk-averse farmer can choose both dosage and timing of application, the total quantity used may decrease under uncertainty. However, if dosage is restricted to the pre-determined label dose, he agrees that use will increase. Auld and Tisdell (1987) extended this analysis and considered that risk-averse farmers may be heterogeneous in their decision criteria. Different farmers may estimate the threshold at different pest population densities and thus make different treatment decisions under the same conditions, depending on whether they want to maximize expected returns or minimize the chance of loss.

Some authors have downplayed the role of farmers' risk preferences and have emphasized the nature of uncertainty in influencing pest management decisions. Plant (1986) examined the same two scenarios as Moffitt (1986): the case where dosage is variable and chosen by the farmer, which he calls the optimizing threshold, and the case where the dosage is set in advance by labeling or regulation and the farmer only chooses whether to apply the pesticide, which he called the discrete choice threshold. The main uncertainty considered here was defined as the variance in the distribution of the potential pest population. Increasing variance has a smaller effect in the discrete choice scenario, but in both cases the expected kill ratio of the pesticide decreases with increasing variation, thus leading to increased pesticide usage. Risk aversion is not a factor in this model. Tisdell (1986) concurred with this assessment, although he described it in terms of the convexity of the functional relationship between the application rate of the pesticide and the response in yield increase. In this system, uncertainty will encourage greater use rates as long as pesticide application has decreasing marginal returns with respect to the yield response. These results are important because they link variations in the optimal use of pest control inputs to the level and nature of uncertainty faced by farmers rather than to differences in their risk preferences.

Farmers' expectations

Since pest population, yield damage, and the relative effectiveness of various pest control practices are stochastic and vary from one location to another and over time, farmers must form effective expectations of the values these variables can take in their fields. Understanding how farmers form such expectations and how consistent those expectations might be with the true underlying conditions is therefore important in pest control. Identifying the sources of variability in pest populations, yield damages, and control effectiveness provides an initial insight. Shea *et al.* (2002) distinguished three types of uncertainty in pest control. There are always questions about the accuracy of any measurements taken, for instance of pest population density, which they call "observation" uncertainty. There is also "model" uncertainty, the lack of complete understanding of the structural relationships in plant–pest–environment interactions. All individual parts in the tripartite model (e.g. environmental conditions, including the weather) are inherently unpredictable in their own right (Gleick, 1987); this "process" uncertainty places strict limits on the accuracy of predictions.[6]

In such an environment of heightened uncertainty, it is unlikely that farmers can form objective expectations. Instead, it is more likely that farmers will form subjective (Bayesian) expectations about the probabilities of future pest conditions and the efficacy of treatments. Generally, subjective probabilities are beliefs held by individuals concerning their degree of uncertainty about any event and the likelihood that it might occur in the future. Formulating subjective probabilities depends on experience, available information, and judgement (Norris and Kramer, 1990). In the context of pest control, different farmers with different experiences, information, or interpretations of events in their fields may look at similar infestations, crop conditions, and weather and develop substantially different expectations of future incidence and severity (see Box 4.1). Individuals can form subjective probability expectations about rare or unique conditions, or even events that have yet to happen (Norris and Kramer, 1990).

Once subjective expectations are formed, they are not fixed; changes in an individual's experience base and available information can allow them to improve their judgement and update their expectations over time, resulting in smaller discrepancies between expectations and actual future conditions

---

Box 4.1. Defining disease incidence and severity.

Plant pathologists use different measures to assess the impact of disease on plant populations. They define incidence as a measure of how many or what proportion of individual plants (or plant parts, depending on the purpose of the study) are involved with the disease in question. Severity, in contrast, pertains only to those plants that are diseased, and reflects the proportion of the individual plant that is involved with the disease in question (Madden and Hughes, 1995; Agrios, 2005). In addition, yield loss is also an important measurement, defined as the difference between actual yield and potential yield that is due to disease (Agrios, 2005).

While plant pathologists are interested in the biological and physiological effect of disease on particular plants, when we discuss soybean diseases and disease management from an economic perspective, a broader view is necessary. For the purposes of this book, we define disease incidence as the frequency at which disease is present in a farmer's fields. We also use disease severity to mean the proportion of a farmer's fields in which the disease is present in any given year. Finally, we define yield loss from a disease as the amount of production that is lost per acre when the disease is present. Differentiating between incidence and severity in such a way is important because their relative size may imply that different decisions for controlling diseases at the farm level might be optimal.

(Pingali and Carlson, 1985). Thus, farmers engage in a Bayesian updating process, where the unique constellation of crop, pathogen, and environmental factors and each season's disease experience add to farmers' information base and are used to refine later predictions (e.g. Keren *et al.*, 2015). Forming and updating subjective expectations can be especially valuable in the case of diseases with little history. Here, any improvement in the information base can have a large positive effect (Yuen and Hughes, 2002).

With subjective expectations about the pest population, N, in their fields and the damage, $D(N)$, that this population might cause, farmers must choose optimal pest control practices, their levels, and timing of application throughout the growing season. Measurement, model, and process uncertainties all impair farmers' learning and the speed of convergence of the posterior distributions to the true underlying likelihoods for N and $D(N)$. The same sorts of uncertainties can cloud farmers' subjective expectation formation and learning when the potential effectiveness of alternative disease management practices, as represented in the kill function $h(N_0,X,A)$, must be grasped. Noisy signals can, once again, affect learning and posterior convergence. New experiences and information used to update subjective probabilities are gathered by farmers throughout each season and from one season to another.

Farmers formulate and update all of these expectations whether they are considering responsive or preventive disease management. As uncertainty mounts, farmers may have stronger incentives to employ preventive management methods (Stern, 1973). Errors in judgement are also costly, but different types of errors may impose different costs on the farmer and prompt distinct responses. For instance, the cost of preventively treating disease in a year when it would not have been necessary is typically relatively small compared with the yield loss from not treating a serious disease outbreak. As the value of the crop increases relative to the cost of the disease management technology, the farmer may have a stronger incentive to employ preventive measures (McRoberts *et al.*, 2011).

It is difficult to overestimate the complexity of the farmer's task of gathering scarce information, formulating subjective expectations, and making pest management decisions. Walls *et al.* (2016) gave an indication in their description of the development, validation, and operation of a computer-based decision support system (DSS) to assist wheat growers in managing barley yellow dwarf virus. This DSS evaluates 72,387 different combinations of input variables in order to generate nine management recommendations. With all this, the DSS is as yet only applicable to a relatively small area – regional models of disease progression are needed for areas with different geographies and environments. Even at this level of complexity and computing power, expert systems are still challenged to provide significantly better recommendations than the farmer's judgement (Gent *et al.*, 2013). It is not only expert systems that may fail to improve upon the farmer's judgement, however. For instance, Szmedra *et al.* (1990) evaluated pest management regimes and found that preventive control based on farmers' experience generally gave greater returns than extension service recommendations based on set population thresholds. They attributed this outcome to the inability of extant methods of calculating threshold levels to adequately deal with stochastic environmental and pesticide efficacy variables. More than highlighting the inherent uncertainties, however, studies like those of Gent *et al.* (2013) may suggest that farmers can form subjective expectations that are consistent and effective in managing pest control, in spite of uncertainty.

Disease Damage Abatement in Soybean Production

The above economic framework of farmers' decision making applies to all manner of crop pests, including pathogens. In the context of managing soybean disease, especially seedling disease and mid-season root rots, the economic considerations of farmers are especially challenging. As described in Chapter 3 (this volume), multiple pathogens – both fungi and oomycetes – can produce the same symptoms and disease progression. The main noticeable symptom in seedling disease is the disease outcome of damping off. At the point when symptoms are visible, it may be too late to take any responsive action, so the primary management option available to farmers may only be prevention. However, as also described in Chapter 3, basic physiological differences between fungi and oomycetes mean that different types of chemical pesticides, usually applied as seed treatments, are sometimes required to combat the different pathogens. Thus, farmers must formulate expectations before planting, usually at the time of seed purchase, as to which pathogens might be present, whether they might become virulent and to what degree, and which of their fields might be affected in the coming season. The potential impact of environmental conditions must also be accounted for. As described in Chapter 2 (this volume), soil moisture is a requirement for pathogen germination and infection, and fungi and oomycetes thrive in lower and higher levels, respectively, of soil water content. Thus, future weather and rainfall patterns, which are inherently unpredictable, will ultimately condition the effectiveness of disease management decisions made before the production season begins.

The situation is much the same with mid-season rots. As the infection enters through the roots, the disease is typically in an advanced stage by the time symptoms reach the stems and become visible. In some cases, the disease can be treated with a systemic fungicide, but for the most part prevention is the primary option. Again, as described in Chapter 2, mid-season rots can be caused by either fungi or oomycetes and must be managed with measures appropriate to the specific pathogen, most often genetic resistance in the seed. Thus, the farmer must develop expectations as to the incidence and severity of a particular pathogen and make management decisions well in advance, usually at the time of seed purchase.

Genetic resistance traits can be effective against both seedling disease and mid-season rots, but only within limits. As described in Chapter 3 (this volume), no traits conferring significant resistance to *Pythium* spp. infection are commercially available. *Phytophthora sojae* resistance traits are more broadly available, but complete resistance is effective only against specific pathogen races that match the *Rps* genes in a given cultivar. The effectiveness of agronomic measures also varies according to the pathogen involved. For example, *P. sojae* is host specific, generally only virulent to soybeans, so crop rotation can be effective in decreasing the pathogen population over time. *Pythium* spp., in contrast, tend to be virulent to a broader range of plant species, often including maize as well as soybeans, so rotation has less impact on this pathogen population.

Using past experience and other information, farmers must formulate expectations about the pathogens and their populations, the potential yield damage they might cause, and the effectiveness of alternative control practices and their timing. How closely such expectations reflect actual conditions determines much about the economics of disease control in soybean production. In the next two chapters, we explore the consistency of subjective expectations formed by US soybean farmers, as well as their influence on disease control strategies.

Summary

The farmer's essential disease management task is to determine whether the yield loss from disease in any given field and year will surpass a threshold known as the EIL and therefore whether it will be cost-effective to prevent or treat the disease. Coming to this decision is the subject matter of the damage abatement model. Subjective expectations held by farmers about future conditions, including pathogen populations and the damage they cause or the relative effectiveness and cost of various practices that may prevent that damage, guide the decision. Measurement errors, structural uncertainties, and inherent unpredictability in key decision variables make the formulation and updating of subjective expectations a challenge for farmers. Negotiating this thicket of uncertainty is therefore the essential challenge in the economics of disease control in soybeans.

Note

[6] In the context of disease management, as described in the previous chapter, the details of the tripartite interaction of crop, pathogen, and environment can have strong effects on the incidence and severity of crop diseases. Yet scientists are only beginning to understand how the soil microbial profile, some members of which have still not been formally described, can play a significant role in disease experience (e.g. Garbeva *et al.*, 2004; Chaparro *et al.*, 2012).

5 Expectations of Incidence, Severity, and Yield Loss

Farmers form subjective expectations of the potential presence of pathogens and their likely impacts on their crops based on personal experiences in years past, observation of conditions in their fields throughout the growing season, experiences shared by neighboring farmers, and information from consultants, extension agents, and input suppliers, as well as from secondary information and data. They also form subjective expectations about how well alternative disease control practices may work on their farms using similar information and means of learning. These expectations, collectively, determine how diseases are managed on individual fields. When farmers' expectations diverge from actual conditions, non-optimal disease management decisions can result. Because little is known about farmers' expectations in soybean disease control, we conducted multiple surveys to discover them. These surveys were carried out at the Economics and Management of Agrobiotechnology Center (EMAC) at the University of Missouri. We examined farmers' expectations of the potential presence and impact of disease, as discussed in this chapter, and of the relative effectiveness of alternative disease management practices, discussed in the following chapter. Once again, we focus our discussion on seedling disease and mid-season root rot, specifically PRR, in order to gain sufficient detail and insight.

Because soybean disease may not occur in every year and in every field, we began by measuring farmers' perceptions of the following:

- *Disease incidence*: the frequency at which diseases are present in their fields.
- *Disease severity*: the proportion of their fields that is likely to be affected when the disease is present.
- *Yield loss*: the amount of production that is likely to be lost when the disease is present.

Together, these three indicators provide a detailed view of the expected damage from disease in a farmer's fields. Using the same indicators, we also evaluated the consistency of farmers' expectations against alternative measures of the true underlying conditions.

Farmer and Expert Surveys

While it is important to assess the reliability of farmers' expectations of disease incidence and severity in their fields, there is no straightforward way of doing so. In most instances, there are no objective measures of disease incidence and severity at the field level, or even at the farm or regional level. For this reason, in addition to surveying soybean farmers for their perceptions of the incidence and severity of diseases, we also surveyed Certified Crop Advisers (CCAs) and used their perceptions as a baseline. Crop consultants are trained agronomists and perform field evaluations through visual inspections and various diagnostic tools across large agricultural production areas. As such, they are in a good position to assess actual soybean disease conditions in any given year and over time.

We administered multiple surveys and surveyed the two respondent groups in different years according to the schedule shown in Table 5.1. Each group was surveyed twice, with 3 years between surveys. Conducting multiple surveys over multiple years allowed us to make comparisons more confidently and to account for temporal variations. All surveys differed in various ways, but care was taken that key questions remained constant across survey instruments in order to allow comparability. In the following sections, we describe the surveys and their findings.

CCA surveys

CCAs are professionals who advise farmers on agronomic practices. The certification was established in 1992 by the American Society of Agronomy (ASA)

Table 5.1. Survey schedule.

Year	Farmer survey	CCA survey
2011		X[a]
2012		X
2013	X	
2014		
2015		X
2016	X	

[a]The 2011 survey was a pilot study and is not used in the empirical analysis.

to provide a benchmark for practicing agronomy professionals in the USA and Canada. In order to become a CCA, one must pass two comprehensive examinations and have a track record of agronomic experience, defined as a combination of formal education and work experience. Once certified, the CCA must document 40 hours of continuing education every 2 years. Most CCAs advise multiple farmers covering tens of thousands of acres. Accordingly, CCAs tend to have both a degree of expertise in crop diseases, broadly, as well as expertise in local and regional conditions, making them particularly insightful on the issues that farmers face.

Crop advisers from 18 major US soybean-producing states were included in the sample. In particular, advisers from 12 states in the north-central region of the USA (Illinois, Indiana, Iowa, Kansas, Michigan, Minnesota, Missouri, Nebraska, North Dakota, Ohio, South Dakota, and Wisconsin) and six states in the south (Arkansas, Kentucky, Louisiana, Mississippi, North Carolina, and Tennessee) were surveyed.

In 2011, an initial trial survey was sent by e-mail to CCAs in order to test and refine the questions that would be used in subsequent surveys. Due to its trial nature, the results of this survey are not emphasized, but relevant results were used for consistency checks. The following year, 2012, and again in 2015, surveys were sent out to the entire population of registered CCAs (6831 in 2015). These surveys were sent via the ASA and had a response rate of 14% in both years. There were 1110 usable responses to the 2012 survey and 1013 in 2015.[7] The general characteristics of the CCA respondents, reported in Table 5.2, demonstrate that the sampled advisers were actively practicing professionals with considerable experience, both in terms of tenure and numbers of acres and farms that they advised. Overall, there was a high degree of consistency in the CCA responses between years.

A primary purpose of the surveys was to record crop advisers' perceptions of the incidence and severity of seedling and mid-season soybean diseases in the USA. The questions focused on the prevalence of stand establishment problems, the frequency with which specific diseases were encountered,

Table 5.2. CCA survey respondent characteristics. (Data from EMAC CCA surveys in 2012 and 2015.)

	Proportion of respondents (%)[a]	
Characteristic	2012	2015
Employed by private firms	62	64
Self-employed	15	20
More than 10 years' experience	70	66
More than 20 years' experience	32	32
Advised on <10,000 acres	37	43
Advised on 10,000–25,000 acres	23	28
Advised on 25-50,000 acres	18	12
Advised on >50,000 acres	22	18

[a]The number of completed responses was 1110 in 2012 and 1013 in 2015.

the capacity to diagnose disease, and the degree to which stand establishment problems have been attributable to seed rot and seedling disease. Midseason questions centered primarily on PRR and its prevalence. Questions on expected damage, as well as the effectiveness of alternative management practices, were also posed with regard to both seedling disease and PRR.

Farmer surveys

Similar surveys were designed and conducted with participating soybean farmers across the USA. The first survey was carried out in 2013 by phone, resulting in a nationally representative sample of 500 soybean farmers.[8] The survey collected information on: farm and farmer characteristics; farmers' perceptions of the incidence and severity of seedling disease and PRR in their fields and regions; farmers' perceptions of the impacts on yields and costs, including replanting and other activities; control practices in use; perceived effectiveness of alternative control practices; and other relevant information, such as the use of insurance. The farmer survey instrument was developed with some questions in common with the expert surveys to evaluate similarities and differences in the perceptions of farmers and experts.

A second farmer survey was conducted in 2016, via the internet, using a shortened version of the 2013 farmer survey instrument. The 2016 survey focused on the same topics as in 2013 with a few omissions, such as those questions dealing with crop insurance and replanting decisions. Here, 479 soybean farmers provided complete responses. Care was taken to ensure that the population sampled was similar to the previous survey and was nationally representative. Table 5.3 reports some relevant characteristics of the respondents to the two surveys. In both surveys, the average producer had been farming for 31 years with between 600 and 716 acres of soybeans. Including other crops (e.g. maize, wheat, cotton) in 2013, the producers farmed an average of 1666 acres, with 48% also having some livestock. Our sample was therefore composed of commercial farmers with sufficient financial means to test their fields for pathogens as well as with access to alternative sources of information that could be used to form adequate expectations of disease incidence and severity.

Table 5.3. Farmer survey respondent characteristics. (Data from EMAC farmer surveys in 2013 and 2016.)

	Average response[a]	
Characteristic	2013	2016
Soybean growing experience (years)	31	31
Soybean planted area (acres)	600	716
Soybean yield (bushels/acre)	44	52
Acres in conservation tillage (%)	71	81
Soybean planting date	5 May	2 May

[a]The number of completed responses was 500 in 2013 and 479 in 2016.

Perceived incidence and severity of seedling disease

Using the 2016 farmer survey, we examined, in some detail, farmers' perceptions of the incidence and severity of seedling disease in fields with stand establishment problems, a key symptom of the disease in soybean production.[9] From the sample of 479 farmers, 177 (37%) indicated that they had experienced stand establishment problems due to seed rot and seedling disease in at least one of the 10 years prior to 2016. However, the incidence of stand establishment problems due to seedling disease was infrequent. Of the 177 farmers who reported stand establishment problems due to seedling disease on their farm, 32% suffered such problems in only 1 out of 10 years, 35% in 2 out of 10 years, 16.5% in 3 out of 10 years, and the remaining 16.5% in 4 years or more (Fig. 5.1).

Figure 5.2 shows that the average incidence of seedling disease varied across states, with northern states having a more consistent and high level of incidence. As is commonly documented, states with heavier soils (e.g. Illinois, Indiana, and Ohio) had especially high incidence. Southern states were less consistent, perhaps due to their fewer and less homogenous soybean operations.

Not all of the acreage on any given farm was affected by the disease in years when it was present. Farmers were asked to indicate the average share of their acres that exhibited stand establishment problems when seed rot and seedling disease were present, and their responses are illustrated in Fig. 5.3. Some 19% of the affected farmers experienced stand establishment problems from the disease on 5% or less of their acres, while another 34% experienced such problems on an average of 6–10% of their acres. Hence, for three-quarters of the farmers who experienced stand establishment problems due to seed rot and seedling diseases on their farm, only 20% or less of their acreage was affected,

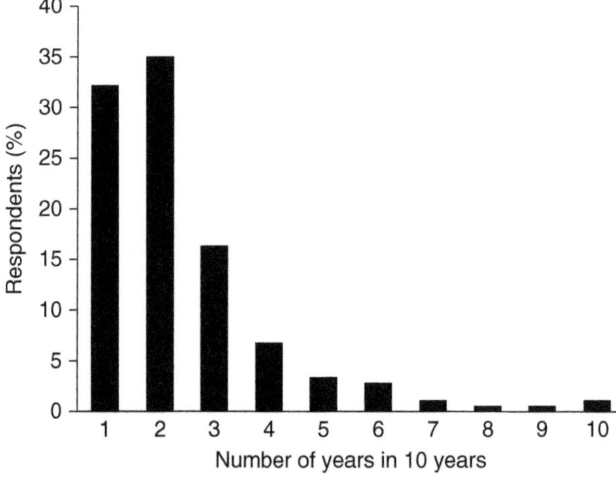

Fig. 5.1. Farmer-reported seedling disease incidence. (Data from EMAC farmer survey in 2016.)

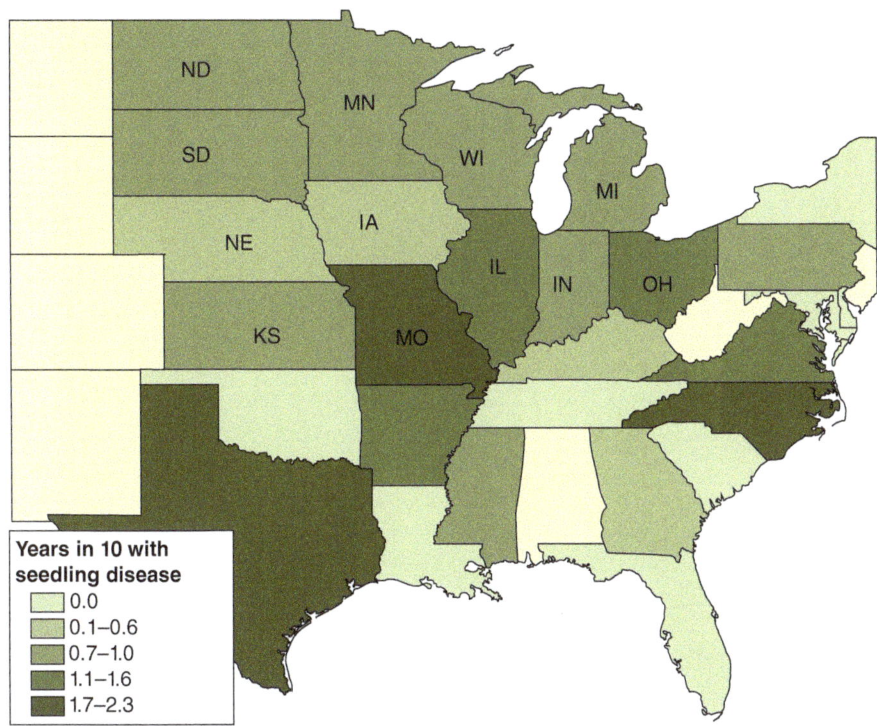

Fig. 5.2. Farmer-reported seedling disease incidence by state. (Data from EMAC farmer survey in 2016.)

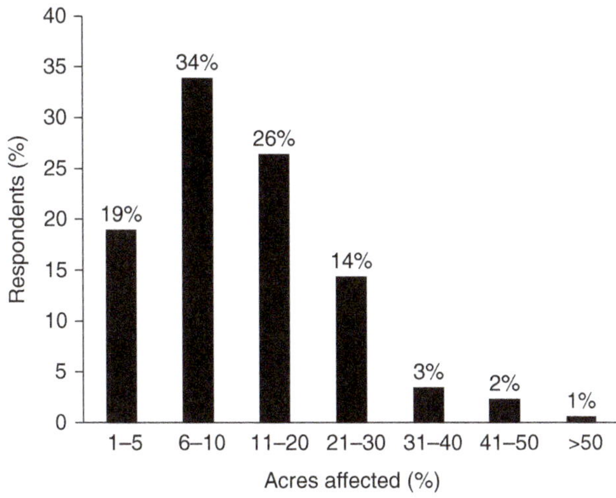

Fig. 5.3. Farmer-reported seedling disease severity. (Data from EMAC farmer survey in 2016.)

on average. A very small percentage of farmers experienced problems on larger parts of their acreage. Geographically, the share of acres affected tended to be relatively consistent across the northern states but much less so in the south, as shown in Fig. 5.4.

Overall, the data provided by the soybean farmers in our sample painted a picture of a disease that affects some farmers, in some years, and on some of their acres. Overall, 37% of the farmers (growing 34% of the soybean acres) perceive seedling disease to occur on their farm and cause stand establishment problems in roughly 1 out of 5 years.[10] On average, 16% of acres experience such problems when the disease was present in a given year.[11] Extrapolating from such incidence and severity figures to national totals would imply that almost 6.9 million acres of soybeans could be affected by seedling disease in any given year, with roughly 1.1 million acres actually experiencing stand establishment problems from the disease.[12]

Using the results of the 2012 and 2015 CCA surveys, we also inferred the severity of stand establishment problems due to seed rot and seedling disease in the USA, as perceived by professional agronomists.[13] The CCAs offered estimates of the share of soybean acres experiencing stand establishment problems from all possible causes (Fig. 5.5), as well as the proportion of stand establishment

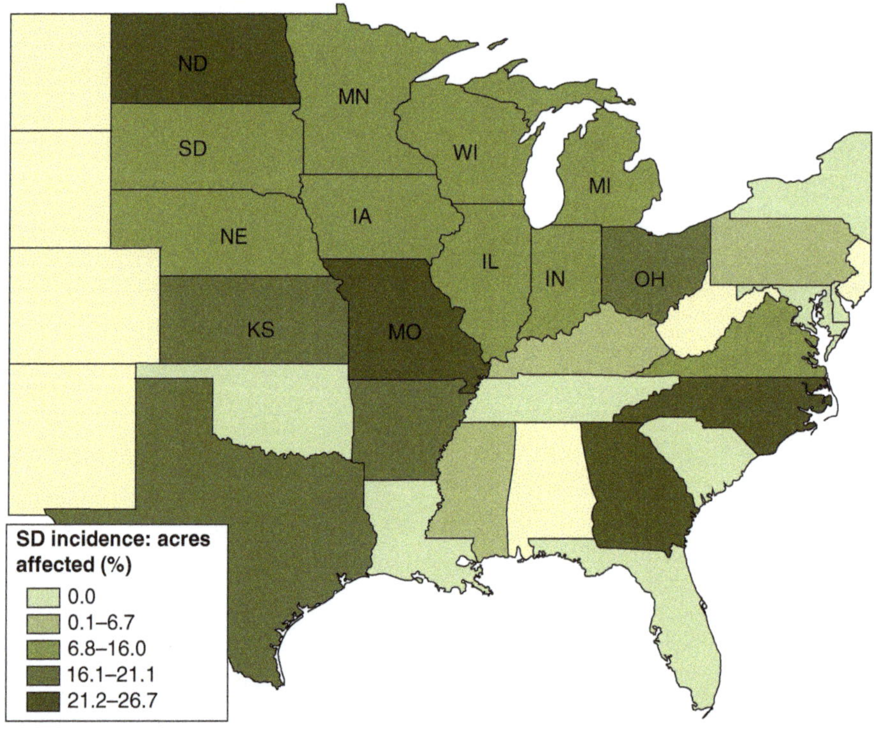

Fig. 5.4. Farmer-reported seedling disease (SD) severity by state. (Data from EMAC farmer survey in 2016.)

problems due to seedling disease (Fig. 5.6). As CCAs experience different levels of disease depending on their location, their individual perceptions are expected to vary from one region to another. However, the distribution of such perceptions across our national sample, as shown in the two graphs, is expected to give a relatively accurate view of disease severity and associated stand establishment problems in the USA. The distributions were very similar in these 2 years and implied that, on average, approximately 9–10.5% of the acres were perceived by the CCAs to suffer some stand establishment problems, of which 16.9–20.2% were due to seedling disease.

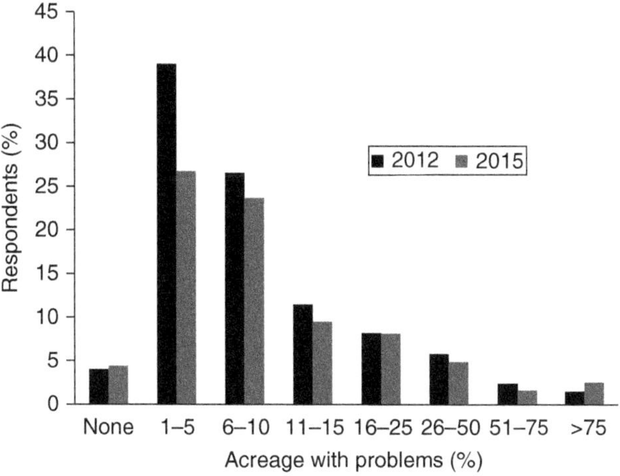

Fig. 5.5. CCA-reported acreage with stand establishment problems. (Data from EMAC CCA surveys in 2012 and 2015.)

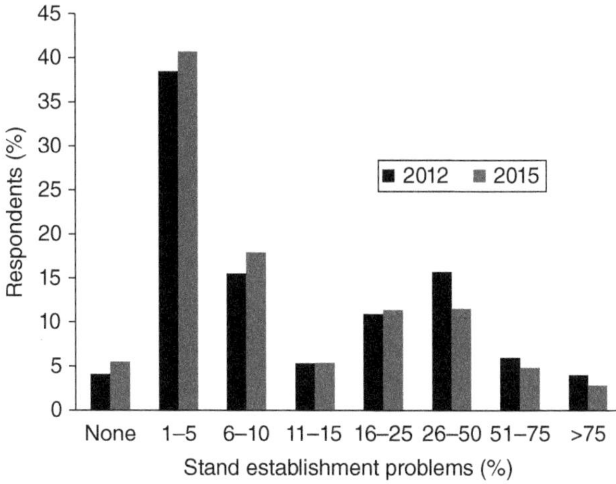

Fig. 5.6. CCA-reported stand establishment problems due to seedling disease. (Data from EMAC CCA surveys in 2012 and 2015.)

Extrapolating from such figures to national totals would imply that an average of 6.9–8.7 million acres of soybeans experience stand establishment problems, of which roughly 1.4–1.5 million acres are due to seed rot and seedling disease.[14] Hence, the perceived severity of stand establishment problems due to seedling disease described by CCAs was similar to that perceived by soybean farmers on their farms.

Perceived yield loss from seedling disease

Both farmers and CCAs were surveyed for their perceptions of yield losses due to seed rot and seedling disease. Evaluating such perceptions is challenging because of the inherent year-to-year variation in the incidence and severity of the disease. For this reason, in the farmer (CCA) surveys, the question was posed as follows: "If seedling disease was completely eradicated in the areas that could be affected on your farm (region), by how much would soybean yields increase on the affected acres, if any?" The responses of the farmers and CCAs are reported in Table 5.4. Farmers' perceptions of yield loss from seedling disease followed a similar distribution in both 2013 and 2016, and averaged 5.77 bushels/acre and 8.2 bushels/acre, respectively, in these 2 years. These estimates are similar, and the higher perceived damage in 2016 might be, in part, attributable to the higher yields obtained by US farmers in that year (on average 52 bushels/acre versus 44 bushels/acre in 2013). CCAs also had similar perceived yield losses on acres affected by seedling disease, averaging 6.8 bushels/acre in 2012 and 7.8 bushels/acre in 2015.

When the farmer-perceived yield loss from seed rot and seedling disease is applied to the estimated areas of 6.9 million acres that could be affected by the disease in any given year, the estimated total loss from seed rot and seedling disease is 47 million bushels of soybeans per year. This estimate of total production loss is consistent with the 44.7 million bushels estimated by the US plant pathologist network to have been lost to the disease each year for the 2006–2014 period (see Chapter 2, this volume, and Fig. 2.5).

Data from both the farmer and the CCA surveys provided additional information on whether farmers were knowledgeable about the pathogens that cause seedling disease in their fields. More than 55% of the farmers indicated that in their most recent encounter with seed rot and seedling disease, the pathogens responsible were identified. This was sometimes done by the farmers,

Table 5.4. Survey question on the expected yield increase if mid-season root rot/seedling disease were completely eradicated. (Data from EMAC CCA surveys in 2012 and 2015 and farmer surveys in 2013 and 2016.)

	CCAs		Farmers	
	2012	2015	2013	2016
Mid-season root rot (bushels/acre)	4.8	4.6	4.9	7.9
Seedling disease (bushels/acre)	6.8	7.8	5.8	8.2

but in other cases through inspections or laboratory testing. Similarly, 70% of the CCAs indicated that they identified the pathogens when they encountered the disease. In terms of the specific pathogens present in diseased seedlings, as Table 5.5 indicates, CCAs and farmers reported very similar experiences in terms of which pathogens were identified, and especially their rankings of the most prevalent pathogens. In addition, the CCAs' responses were very consistent across both surveys. *Pythium* and *Phytophthora* spp. were the most commonly reported pathogens, present in 70–80% of seedling disease events in nearly all cases, which is supported by field tests conducted by Rojas *et al.* (2017a). *Rhizoctonia* and *Fusarium* spp. were less common, but likewise found in similar numbers across the three surveys. Other pathogens are found in a relatively small portion of seedling disease cases. Overall, these results provide evidence that US soybean farmers, as a group, form effective expectations of the incidence and severity of seedling disease on their farms, which are consistent with those of experts and with reported aggregate measures.

Overall, we found that the perceptions of US soybean farmers of the incidence, severity, and yield losses from seed rot and seedling diseases affecting their fields, as well of the pathogens that cause them, were consistent with national statistics provided by experts, such as CCAs and state plant pathologists. We therefore conclude that farmers, as a group, form effective subjective expectations about the incidence, severity, and damage in the case of seedling disease, and we expect these expectations to be in line with actual occurrences.

Perceived incidence, severity, and yield loss from PRR

Using the 2016 farmer survey, we also examined farmers' perceptions of the incidence and severity of PRR.[15] From the sample of 479 soybean farmers, 162 (34%) indicated that they had experienced problems with PRR in at least 1 of the 10 years prior to 2016. However, the incidence of PRR was infrequent for most farmers. Of the 162 farmers who experienced PRR on their farm, 35% had problems in 1 out of 10 years, 38% in 2 out of 10 years, 15% in 3 out of

Table 5.5. Seedling disease experience. (Data from EMAC CCA surveys in 2012 and 2015 and farmer surveys in 2016.)

	Farmers	CCAs	
	2016	2012	2015
Question: "Thinking about your most recent encounter with seedling diseases and/or seed rot, was the causal agent identified?"			
Answer: yes (%)	56	70	69
Question: "What pathogens were found present?"			
Answer: response (%)			
Pythium	53	78	79
Phytophthora	73	76	79
Rhizoctonia	34	55	60
Fusarium	32	40	51
Other	5	6	10

10 years, and the remaining 12% in 4 or more years (Fig. 5.7). Geographically, the incidence generally was consistently higher in the northern states, as shown in Fig. 5.8. North and South Dakota tended have high incidence as well as the states known for heavier soils (Illinois, Indiana and Ohio). Other states, especially southern states, showed more variation.

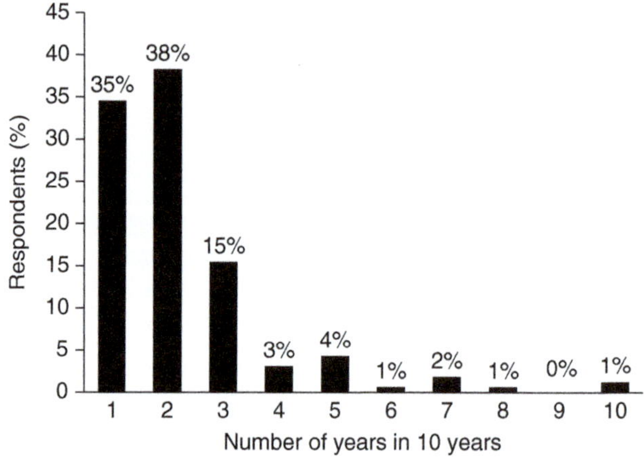

Fig. 5.7. Farmer-reported PRR incidence. (Data from EMAC farmer survey in 2016.)

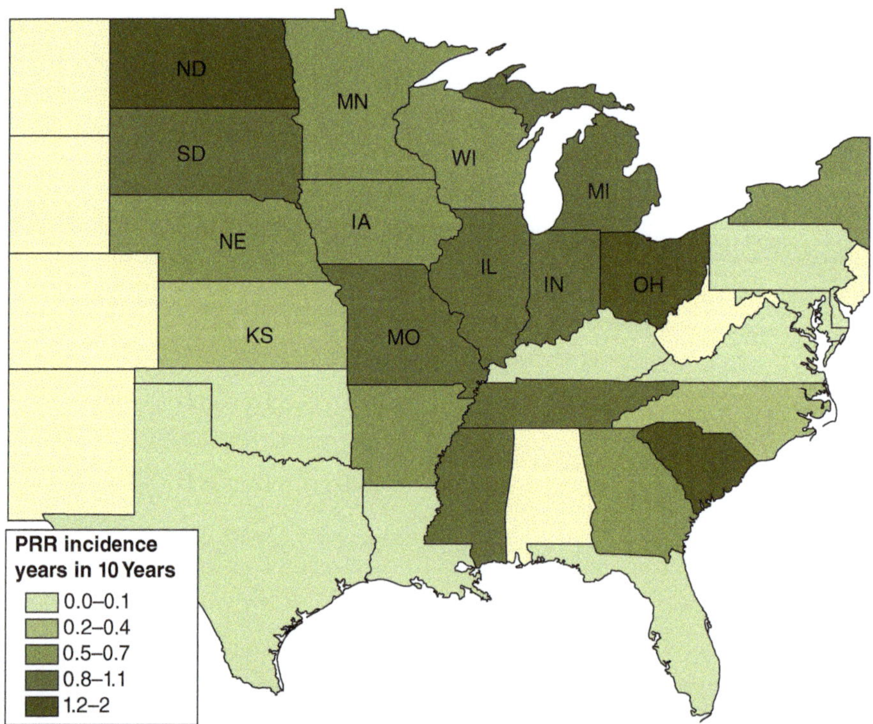

Fig. 5.8. Farmer-reported PRR incidence by state. (Data from EMAC farmer survey in 2016.)

As in the case of seedling disease, even in the years when PRR was present on their farms, not all of the farmers' fields and acres were affected. As Fig. 5.9 illustrates, on average, 30% of the farmers experienced problems from PRR on 5% or less of their acres, while another 36% experienced such problems on an average of 6–10% of their acres. Hence, for more than two-thirds of the farmers who experienced PRR on their farm, only 1 out of 10 acres or less was affected when the disease was present. Another 18% of the soybean farmers experienced PRR problems on up to 20% of their acreage and 17% of farmers experienced problems on more than 20% of their acreage when the disease was present. Interestingly, the responses showed little geographical patterning other than a generally higher share of acres affected in northern states (Fig. 5.10).

The data provided by the soybean farmers in our sample once again paints a picture of a disease that affects some farmers, in some years, and on some of their acres. Based on these data, we can infer that about one-third of all farmers (and an equal share of soybean acres) perceive PRR to occur on their farm and cause noticeable problems in roughly 1 out of 5 years.[16] On average, 13.7% of the acres experienced such problems in the years when PRR was present.[17] Extrapolating from such figures to national totals would imply that an average of almost 6.5 million acres of soybeans could be affected in any given year by PRR, of which roughly 0.88 million acres actually experience root rot problems.[18]

Using the results of the 2012 and 2015 CCA surveys, we also solicited the perceptions of these professional agronomists about the severity of PRR. The CCAs estimated that, on average, 9.7% of the acres could be affected by PRR in any given year. Extrapolating from these figures to national totals would imply that an average of 7.5–8 million acres of soybeans could be affected in any given year by PRR, to some degree[19]. These acreage totals are somewhat higher but within range of the 6.5 million acres projected from the farmer perceptions above.

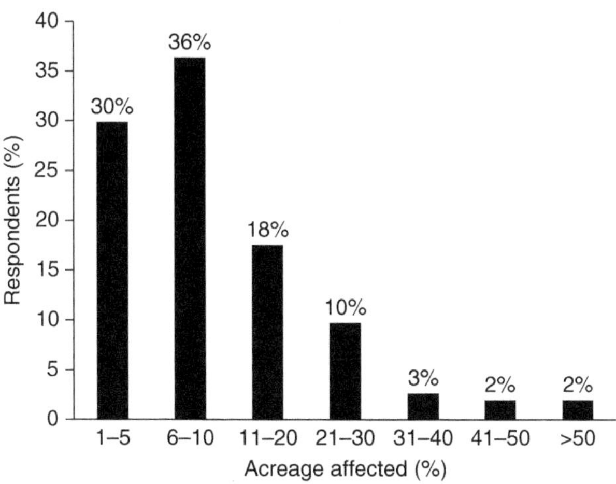

Fig. 5.9. Farmer-reported PRR severity. (Data from EMAC farmer survey in 2016.)

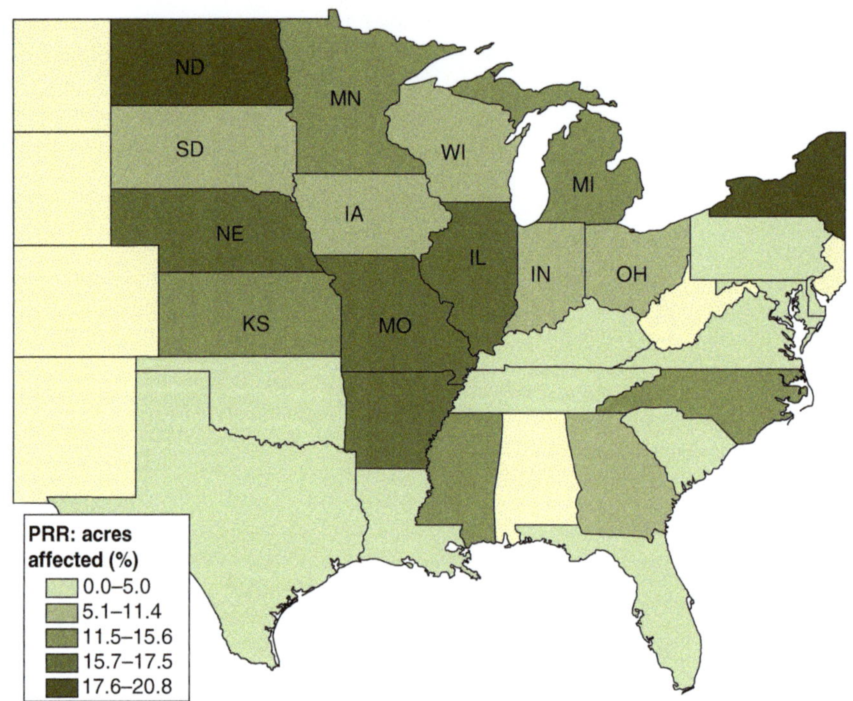

Fig. 5.10. Farmer-reported PRR severity by state. (Data from EMAC farmer survey in 2016.)

Both farmers and CCAs were also surveyed for their perceptions of yield losses from PRR. Because of the inherent year-to-year variation in the incidence of the disease, the farmers (CCAs) were asked the following question in the survey: "If PRR was completely eradicated in the affected areas of your farm (region), by how much would soybean yields increase on the affected acres, if any?" The responses of the farmers and CCAs are reported in Table 5.4. Farmer perceptions of damage from PRR averaged 4.9 and 7.9 bushels/acre, in 2013 and 2016, respectively. These estimates are similar, and the higher perceived damage in 2016 might be, in part, attributable to the higher national yield average for that particular year (52 versus 44 bushels/acre in 2013). The CCAs also had similar perceived yield losses from PRR in affected areas, averaging 4.8 bushels/acre in 2012 and 4.6 bushels/acre in 2015.

When the farmer average perceived yield loss from PRR is applied to the estimated area of 6.5 million acres that could be affected by the disease in the USA in any given year, the estimated total production loss from PRR is 41.1 million bushels of soybeans per year. When the CCA average perceived damage from PRR is applied to the estimated 7.5–8 million acres that could be affected by PRR, the estimated total production loss is 36.8 million bushels of soybeans per year. The estimates of total production loss derived from the perceptions of both farmers and CCAs are also similar to the average annual estimated loss of 36.7 million bushels calculated by the US plant pathologist network for the 2006–2014 period (Bradley *et al.*, 2016).

Overall, we found that US soybean farmers formed subjective expectations about the incidence, severity, and yield loss associated with PRR affecting their fields that were consistent with national statistics provided by experts, such as CCAs and state plant pathologists. We therefore conclude that farmers form effective subjective expectations about the incidence and severity of PRR, and we expect these expectations to be in line with actual occurrences.

Summary

For the most part, there are no objective measures of disease incidence or severity available to farmers. Thus, farmers must form their own subjective expectations about the likely incidence and severity of disease in their fields, as well as what level of yield loss, if any, they might experience in any given year. Such expectations are the foundation of the disease management decisions that they must make every growing season. The lack of objective measures also makes it difficult to assess the accuracy of farmers' expectations. In order to discover and quantify farmer expectations, we surveyed national samples of soybean farmers in different years. In order to evaluate their accuracy, we also surveyed national samples of CCAs. Like most soybean diseases, seedling disease and mid-season rots are infrequent and affect only some of the soybean farmers in the USA. On average, roughly one-third of all soybean farmers experience a problem once every 5 years and on only a portion of their acreage. Nevertheless, when these rates are extrapolated to the national level, the value of the loss can be quite large. We also found broad agreement among reported incidence, severity, and yield loss among farmers, CCAs, and state plant pathologists, which lends strong credence to the accuracy of farmers' subjective expectations and disease management decisions.

Notes

[7] Detailed descriptions of the sampling procedures, the survey instrument, and summary statistics are reported by Arbuckle *et al.* (2012, 2015).

[8] Detailed descriptions of the sampling procedures, the survey instrument, and summary statistics are reported by Kalaitzandonakes and Kaufman (2013).

[9] The 2013 farmer survey examined, in detail, stand establishment problems, both from seedling disease and in general. Findings from the first survey were used to refine the survey instrument used for the second farmer survey, so the figures cannot be fully merged as some questions were not identically stated. Overall, analysis of the two surveys yielded qualitatively similar results; for simplicity, only the 2016 figures are presented here.

[10] The sample average for the 177 farmers who experienced stand establishment problems due to seedling disease was 2.42 out of 10 years.

[11] The sample average for the 177 farmers who experienced stand establishment problems due to seedling disease was 15.9% of their acres having such problems in years when the disease was present.

[12] Using 2016 national acreage figures, extrapolation from our figures implies that out of 83.5 million acres planted, 34% acres could be affected at a frequency of 0.242; hence, a total

of 6.87 million acres could be affected in an average year. Overall, 15.9% of these acres, on average, would experience stand establishment problems, or 1.09 million acres. Varying the planted acreage from year to year would produce variations in the acreage that could be affected and experience stand establishment problems.

[13] Because CCAs do not generally evaluate the same acres every year, and most (86%) of them encounter seed rots and seedling disease every year, they were not asked to report on the incidence of the disease.

[14] Using 2012 and 2015 national acreage figures for the two surveys, respectively, extrapolation from our figures implies that out of 77.2 million and 82.6 million acres planted in these two years, 9% and 10.5% of them, respectively, could have stand establishment problems; hence, 6.9–8.7 million acres could show such problems. Applying the average share of such problems attributable to seedling disease in the two years yields an estimated 1.4–1.5 million acres for these two years.

[15] The 2013 farmer survey also examined, in detail, the incidence and severity of PRR. Findings from the first survey were used to refine the survey instrument used in the second farmer survey, so the figures of the two surveys cannot be combined in all instances as some questions were not identically stated. Overall, analysis of the two surveys yielded qualitatively similar results; for simplicity, only the 2016 figures are presented here.

[16] The sample average for the 162 farmers who experienced problems due to PRR was 2.27 out of 10 years.

[17] The sample average for the 162 farmers who experienced problems due to PRR was 13.7% of their acres having such problems in years where the disease was present.

[18] Using 2016 national acreage figures, extrapolation from our figures implies that out of 83.5 million acres planted, 34% of acres could be affected at a frequency of 0.227; hence, 6.44 million acres could be affected. Overall, 13.7% of these acres (0.88 million acres), on average, would experience PRR problems. Varying the planted acreage from year to year would produce variations in the acreage that could be affected and that actually experience PRR problems.

[19] Using 2012 and 2015 national acreage figures for the two surveys respectively, extrapolation from our figures implies that out of 77.2 million and 82.6 million acres planted in these two years, 9.7% – hence 7.48–8.02 million acres – could be affected by PRR.

6 Disease Control Methods and Effectiveness

Farmers decide whether to use preventive and responsive disease control methods on their farm after they have formed subjective expectations about the potential damage from disease in their fields and the potential means to control it. The subjective expectations about the relative effectiveness of alternative disease control methods are, as before, based on: the farmers' past experiences; the experiences of other farmers in the region; information from consultants, extension agents, and input suppliers; and secondary data. When they implement one or more disease control methods, they must integrate them with weed and insect pest management, soil health and fertility, labor and machinery use, and other farm management practices in order to maximize profits (Chaube and Singh, 1991).

As little is known about the practices that soybean farmers use to control disease on their farm or about their expectations of relative effectiveness, we use surveys to infer them. Once again, we focus our discussion on seedling disease and mid-season rots in order to obtain sufficient detail and insight. We begin by reviewing the range of disease management options available to soybean farmers, the main mechanisms by which they control disease, and the factors that condition their effectiveness.

Seedling Disease and Mid-season Root Rot Control Methods

There are three general categories of disease control methods in soybean production: agronomic practices, chemical controls, and genetic resistance in the seed. As discussed previously, disease is not simply the result of a particular pathogen being present in a field but rather the outcome of complex interactions between the pathogen, the host, and the environment. In this context, the effectiveness of all disease control practices can be highly stochastic and confounded by conditions outside the farmer's control. Furthermore, some control methods may be more complex to implement, involving multi-year considerations and use of specialized farm assets, while others may be easier to use within a single growing season and mostly independent from other inputs. Finally, the various disease control methods also tend to have significantly different unit costs.

Agronomic methods

Nearly all of the agronomic decisions that farmers make have some impact on crop disease. The modern practice of planting genetically similar plants in more or less continuous monoculture and in high population densities can contribute to the severity of crop disease. Extensive monoculture provides soil pathogens with a large area of ready hosts, allowing pathogen populations to increase, and exerts selection pressure for individual pathogens well adapted to colonizing that one crop variety (van Bruggen and Termorskuizen, 2003). The incredible productivity that modern agriculture has attained is largely a result of intensive monocultures; most agronomic methods of disease management currently in use work to moderate pathogen-promoting effects of monoculture and retain high levels of production.

In general, agronomic measures involve manipulating the physical environment in which crop plants and pathogens grow and develop, to the benefit of the crop and the detriment of the pathogen. Farmers can do this in three general ways. Crop rotation and sequencing programs vary the potential host profile for soil pathogens and influence soil health and fertility, which contributes to crop vigor and the ability to withstand disease. Manipulating physical soil characteristics, including tillage and cultivation practices, can make the soil more or less hospitable for both crops and pathogens. Finally, there is a range of options surrounding planting, such as timing, spacing, and fertilization, that can promote or inhibit the development of disease. We consider each in turn.

Crop rotation

In general, crop rotation can suppress pathogen populations by introducing non-host plants into an infested field. However, the cropping sequence that is used matters; in some cases, alternative crops can harbor pathogens without developing disease, potentially making disease problems more severe in subsequent years. The same is true for cover crops that might be used to combat soil erosion during the off season. With the broad adoption of inexpensive synthetic fertilizers from around 1950 onwards, the benefit of crop rotation for soil fertility became less certain. As fertilizers alone could maintain soil fertility at very high levels, the option of growing the most profitable crop continuously, or nearly so, became more attractive (Rush *et al.*, 1997). Around 1990, with changes in commodity demand and production, it became more common for US farmers to plant soybeans for 2–5 years consecutively before rotating to maize. This more intensive soybean production created conditions favorable to increased *Phytophthora sojae* populations and may have increased disease incidence and severity (Dorrance *et al.*, 2009).

Since soybean is the sole host plant for *P. sojae*, rotation to other crops can do much to decrease the pathogen population. Depending on the population of *P. sojae* in a particular field, the timing of crop rotation may need to be adjusted to substantially decrease the pathogen population. *P. sojae* produces spores that can survive in the soil for many years in the absence of host plants and throughout periods of abiotic stress, such as drought. The longer the rotation away from soybeans, the more the pathogen life cycle can be broken and the population decreased (Zhang and Xue, 2010). *Pythium* spp., in contrast, are little affected by the standard soybean/maize rotation. Many *Pythium* spp. are virulent on maize as well as soybeans (Zhang and Yang, 2000; Matthiesen *et al.*, 2016; Radmer *et al.*, 2017), so farmers must use different crop rotations or other methods for *Pythium* spp. control.

Tillage, cultivation, and drainage

Historically, tillage has been used to control weed growth, loosen the soil to allow easier crop root growth, and improve drainage. It also leaves the soil surface largely bare and unprotected from wind and water, the primary agents of erosion. Excessive tillage resulted in topsoil loss in the central and southern Great Plains of the USA and was instrumental in creating the conditions that led to the Dust Bowl of the 1930s. In response, farmers and agronomists began to develop various methods of conservation tillage, involving planting a new crop with minimal, or no, soil disturbance. The introduction of genetically engineered HT crops and their associated broad-spectrum herbicides in the 1990s made the weed control effect of tillage less important and accelerated the adoption of conservation tillage in North and South America (van Bruggen and Termorskuizen, 2003).

The choice of tillage system can have mixed effects on disease incidence and severity through several different mechanisms. Mechanically disturbing the soil can break up fungal hyphae and expose them to the sun and weather, but it can also spread spores to new areas. Incorporating crop residues into the soil can make that food source more available to soil pathogens. If residues are buried more deeply, however, they may be beyond the reach of pathogens. Breaking up crop residues may expose pathogen colonies to other, hostile microorganisms. The timing of incorporation also matters. Crop residue left on the surface for a time can be colonized by other saprophytic fungi and thus be unavailable to pathogens after tillage. In conservation tillage systems, where residues remain on the surface long term, the underlying soil is typically cooler and wetter than it otherwise would be, fostering a higher pathogen population. The increased water availability, however, may result in faster and more vigorous crop growth that can overcome the greater disease potential.

Cultivation during the growing season helps control weeds, improves soil aeration, and increases soil permeability to water, all of which improve the crop's growing environment, assisting plant vigor and reducing disease problems. However, the increased machinery traffic can result in damage to crops that can provide a means of entry for pathogens. Conservation tillage systems can provide significant benefits in the form of less soil compaction and cost savings in terms of fuel and machinery use. However, they can also create a soil environment more conducive to oomycete growth and development. Conservation tillage practices have contributed to the increasing incidence of seedling disease since around 2000, as well as an increasing genetic diversity in *P. sojae* populations (Dorrance *et al.*, 2009). The improved drainage achieved by tillage can be important in decreasing disease incidence, as oomycetes require waterlogged soil in order for spores to move to and infect roots. In addition, tillage can bury oospores below the root zone, limiting their ability to colonize soybean roots (Zhang and Xue, 2010).

Improving field drainage by tiling can be a highly effective means of combatting disease. Drainage improvements differ from other agronomic measures in that they are a long-term capital investment rather than a recurring annual expense. However, in some areas, environmental regulations restrict the amount of drainage pipe that can be installed in fields. Furthermore, an ageing drainage infrastructure can lose its effectiveness and must be replaced. Both of these factors limit the feasibility of improving drainage (Dorrance *et al.*, 2009).

Planting strategies

The planting process begins with seed selection; selecting quality seed is important for disease control as well as overall production success, as seed genetics determines the yield potential of any given field. All other things being equal, healthier, more vigorous seedlings are less likely to be adversely affected by diseases (van Bruggen and Termorskuizen, 2003).

One of the many complex decisions farmers have to make is that of planting date. A number of factors enter into this decision, including soil conditions, weather, prospects for crop growth, and the time required for crop maturation. In the context of disease management, the choice of a planting date has one main objective: to minimize the time that a developing seedling is most susceptible to infection by a pathogen (Chaube and Singh, 1991). This is done by planting at a time when soil moisture and temperature are more suitable for crop development than for pathogen activity. The more that crops can grow before soil conditions become optimal for pathogens, the less susceptible to infection they will be. Depending on the specific crops and pathogens involved, this can mean adjusting the planting date to earlier or later, in order to plant in either cooler or warmer soil, possibly at different moisture levels. The goal is the same, but the decision changes depending on the conditions a specific pathogen finds more favorable (Rush *et al.*, 1997). As such, farmers' expectations about the pathogen population profile in their fields can condition their decisions.

Planting density can also influence the incidence of disease. A denser crop stand will generally have a more closed canopy, creating a more humid microclimate under the canopy. This environment is more conducive to foliar disease, although studies suggest that if the general weather conditions are especially favorable or unfavorable to pathogen development, it will swamp the effects of the canopy microclimate (Krupinsky *et al.*, 2002). However, a higher plant density can decrease the severity of many soil-borne diseases, such as those caused by oomycetes. Most soil pathogens are monocyclic, with only one life cycle and thus one infection opportunity in each growing season. With a finite amount of inoculum, a denser stand means more plants may escape infection. Some soil pathogens are polycyclic, however. In these cases, a denser stand would confer no particular advantage in disease management (Rush *et al.*, 1997).

Other agronomic methods

Once the seeds are in the ground, crop nutrition for the season (over and above general soil-quality considerations) comes to the fore. There are two parts to this: fertilizer and water. Synthetic inorganic fertilizers can optimize the levels of major nutrients on nearly any plot of land for most of the major crops. In addition to maximizing crop production, crop nutrition has two major crop protection goals: first, to minimize plant stress and promote healthy, vigorous growth, making the crop less susceptible to disease, and second, to manipulate soil nutrients to the simultaneous advantage of the crop and disadvantage of pathogens.

The main macronutrients contained in synthetic fertilizer are nitrogen, phosphorus, and potassium. Sufficient nitrogen is extremely important for plant growth and health, and a lack of nitrogen can make plants weak and more susceptible to many diseases. Soybeans, as legumes, derive a large part of the nitrogen they need from the atmosphere through their root nodules, but they still take some elemental nitrogen from the soil. Treating soybeans with nitrogen is therefore not common. Research has indicated that, under some circumstances, soybeans can show a yield increase from nitrogen fertilizer application, but it is not clear that this would provide a net economic benefit (Osborne and Riedell, 2006). Added phosphorus can speed maturity and stimulate root activity, improving plant health. However, some phosphorus supplements have been associated with increased pathogen growth and disease severity (Sanogo and Yang, 2001). Higher potassium levels have been shown to directly inhibit pathogen activity and reduce the incidence and severity of a host of fungal and oomycete diseases (Amtmann *et al.*, 2008; Ghorbani *et al.*, 2008). Overall, the potential impact of fertilizer on soybean disease is unclear.

The amount of water present in the soil affects all aspects of crop growth and development. More than 80% of plant tissue is composed of water, making it the single most important substance to an actively growing plant. Adequate water is thus critical to plant health, leading producers of many crops to employ irrigation. Soybeans, due to their extensive root system, have fewer water issues than other crops and are generally managed as a dryland crop. Only about 10.5% of US soybean acreage is irrigated (USDA, 2018). Thus, from a disease management perspective, the main decision point for farmers is in ensuring adequate drainage for natural soil water content. Irrigation management is not a factor considered in this study.

Chemical controls

Chemical fungicides have a long history of use in agriculture. The use of fungicidal seed treatments has been documented as early as the middle of the 17th century. One of the best known fungicides, Bordeaux mixture, was developed in France in the late 19th century to protect grapes from downy mildew (Morton and Staub, 2008). Fungicides constitute some 15% of the market value of all crop protection chemicals (USDA, 2018). Farmers often manage foliar diseases, such as soybean rust, by spraying foliar fungicides during the growing season. Fungicides used to combat seedling disease and root rots in soybeans can be applied directly to the soil or as a seed treatment. Some soybean fungicides are available for in-furrow soil application, but cost and effectiveness issues can make this a less attractive option. Seed treatments provide another mode of application for fungicides used for seedling disease. The coated seeds deliver a dose of fungicide to the soil in the immediate area around the new seedling and can provide a systemic fungicide to the stem and foliage of the developing plant. Chemical seed treatments can also be effective in suppressing seed-borne diseases (Mueller *et al.*, 2013). With either of these methods, the treatment decision is preventive and must be made before planting.

Fungicides used to control seedling disease and root rot in soybeans come from two chemical families. The most commonly used fungicides are metalaxyl and mefenoxam. These are stereoisomers and are both phenylamides (Hamlen *et al.*, 1997). They have the same mode of action, inhibiting RNA synthesis and thus impeding mycelial growth and infection (Mueller *et al.*, 2013). Metalaxyl is available only as a seed treatment (e.g. Allegiance®; Bayer, 2016). Mefenoxam is available as a seed treatment, either as the sole active ingredient (e.g. Apron XL®; Syngenta, 2016b) or in combination with other fungicides to minimize the development of resistance (e.g. ApronMaxx®; Syngenta, 2016a). Mefenoxam is also available in granular form to be applied to the soil either before or after planting (e.g. Ridomil®; Syngenta, 2016c).

Recently, another fungicide targeting oomycetes has been approved for use in soybeans, ethaboxam. Ethaboxam, a thiazole carboxamide, inhibits cellular growth and elongation by disrupting microtubules in oomycete hyphae cells (Uchida *et al.*, 2005). It also appears to have other modes of action that are still under investigation. With the possibility of multiple modes, the incidence of seedling disease pathogen resistance to ethaboxam may be lower (Kim *et al.*, 2004). It is available as a seed treatment in a combined formulation with two other fungicides with different modes of action, in order to further minimize the development of resistant oomycete strains (Valent, 2016).

Genetic resistance

As discussed briefly in Chapter 3 (this volume), disease resistance in soybeans can be classified into two different types: specific and partial. In the case of specific resistance, a soybean cultivar is completely immune to one or more pathogen varieties, or races, but other races can still cause disease in that cultivar.

A partially resistant cultivar displays a certain amount of resistance to all races of a particular pathogen but is not immune to any of them, and all pathogen races can still cause a low level of disease. These two types of resistance also differ in their genetic basis and prospects for maintaining their effectiveness in preventing disease. In both cases, resistance is the result of a lack of compatibility, to some degree, between the host plant and the pathogen. This can occur either when the pathogen cannot chemically recognize the host plant or when the host is able to defend itself, partially or completely, against infection by the pathogen (Boyd *et al.*, 2013). Most crop plants are resistant to infection by nearly all microorganisms with which they might come into contact during the growing season. In only a few cases does a particular species come to recognize a plant as a potential food source to the extent that a pathogenic relationship can develop. This places selection pressure on the host plant. Through natural genetic variability, mutation, or other means, the host population profile changes to include more individuals resistant to the disease. In turn, the pathogen population is pressured to exploit these new host varieties in order to ensure its own survival. Through this process of coevolution, the genetic variability of both populations develops such that they maintain a sort of equilibrium (Fry, 1982, pp. 207–208).

Specific resistance

Specific resistance to a particular pathogen race is typically controlled by a single resistance gene. Even if a cultivar has multiple resistance genes, conferring resistance to multiple pathogen races, these genes operate independently. For this reason, specific resistance is often called monogenic or oligogenic resistance (Schumann and D'Arcy, 2010, p. 217). In most instances of specific resistance, the host resistance gene counters the effects of a specific virulence gene in the pathogen in what is known as "gene-for-gene" interaction. Specific resistance is a qualitative phenomenon: the cultivar either is or is not resistant to the relevant pathogen race.

When the pathogen–host relationship is characterized by gene-for-gene interaction, four different genetic outcomes are possible. The host gene for resistance is usually dominant (R) over the susceptibility allele (r). In the pathogen, by contrast, the gene for avirulence is usually dominant (A) over the virulence gene (a). A pathogen race consists of individuals with one specific A allele out of many possibilities (A_1, A_2, etc.). The specific A gene codes for the production of a particular protein, known as an elicitor, that is recognized by host individuals with the corresponding R gene (R_1, R_2, etc.), which codes for a receptor for that one elicitor. When the elicitor and receptor come together (AR combination), it triggers a defensive response by the host that prevents infection. If the host does not produce a receptor for the specific elicitor (Ar combination), or if the pathogen does not produce an elicitor the host can recognize (aR combination), or if neither the elicitor nor the receptor is present (ar combination), the plant becomes infected and develops the disease (Agrios, 2005, pp. 140–141).

Gene-for-gene interaction has two significant implications for plant breeders and, by extension, for soybean farmers. First, if breeders wish to rely on specific resistance to protect a particular cultivar from disease, they must insert multiple resistance genes into the crop genome. Each cross required to do this, however, holds the potential to introduce other traits that reduce yield or produce other undesirable characteristics. Thus, a considerable amount of time may be necessary to develop a cultivar with a wide range of specific disease resistance. Second, the effective lifespan of each *R* gene is often limited. While there are cases of specific resistance that has been effective for extended periods, if the pathogen population is genetically diverse, the profile can change quickly, with individuals producing a novel elicitor becoming dominant. Such individuals will be virulent against the previously resistant cultivar.

Partial resistance

In some cases, a cultivar is not completely resistant to a disease, but the degree of infection and the severity of disease symptoms are significantly less than with a susceptible crop variety. Partial resistance, sometimes referred to as rate-reducing resistance, is usually the result of multiple genes that interact additively, each counteracting pathogen activity in one small way. Partial resistance is therefore usually characterized as polygenic, in contrast to oligogenic specific resistance (Schumann and D'Arcy, 2010, p. 219). Partial resistance is not race specific, but instead operates with more or less equal effectiveness across all races of a particular pathogen. Thus, it is a quantitative phenomenon: individual cultivars vary in the degree to which partial resistance protects them from disease rather than which pathogen races they are protected from.

Because of the polygenic nature of partial resistance, it is robust against genetic diversity in pathogen populations. The amount of genetic change that would have to come together in one set of individuals is significantly large so that the likelihood of resistance is very low. Also, partial resistance may not operate according to gene-for-gene interaction, further reducing the probability that a pathogen population could evolve to circumvent it. Partial resistance is thus much more durable over time than specific resistance (Schumann and D'Arcy, 2010, p. 219).

Outside factors can condition the effectiveness of partial resistance. Two of these are of greater importance: the size of the pathogen population and the environment. A very large pathogen population, producing large amounts of inoculum, can overwhelm a partially resistant cultivar and cause serious disease. If environmental conditions (e.g. temperature, moisture levels) in a particular year favor pathogen development, it can result in a pathogen population that is larger, more vigorous, or both, and thus can produce more serious disease conditions. There is also some evidence that environmental factors such as light level, day length, and temperature can directly influence partial resistance, producing differing disease severity with no change in the pathogen population (Fry, 1982, pp. 228–230).

Tolerance

The term tolerance is sometimes used to mean partial resistance, but in the strict sense, it is not a form of resistance at all. Instead, tolerance refers to the ability of a plant to withstand damage and minimize impact, especially in the form of lost yield that a cultivar experiences with a particular severity of infection and disease. Thus, whereas resistance is the ability to limit disease severity, tolerance is the ability to produce in spite of disease. Tolerant plants are still susceptible to a disease pathogen but do not die from the disease, and their yield is reduced much less than that of a fully susceptible cultivar. The mechanism behind tolerance is not well understood (Agrios, 2005, p. 139).

Sources of resistance

Because of their genetic uniformity, commercial soybean cultivars generally offer little promise as sources of resistance to new pathogens. Instead, breeders look to exotic germplasm, landraces, or even wild types for the genetic resources they need to combat disease (Carson, 1997). Plant scientists have long been aware of the need for a comprehensive inventory of the world's genetic resources to promote crop quality in general and disease resistance in particular. Some of the most valuable resistant soybean cultivars were derived from germplasm samples collected in the early 20th century (Hymowitz, 1984). Ideally, the area where the crop's wild progenitor and the pathogen originally coevolved, if that can be identified, would offer the greatest amount of potentially novel resistance genes for use in crop breeding programs (Dorrance and Schmitthenner, 2000). Unfortunately, the further removed this germplasm is from currently well-adapted cultivars, the greater potential it has for carrying undesirable agronomic traits that could be transferred along with disease resistance. This could complicate the breeder's task and extend the time necessary to produce new resistant cultivars (Carson, 1997). Novel gene-editing methods can accelerate the introduction of disease resistance in elite lines, and genetic engineering techniques offer an alternative method of transferring resistance genes into crop plants without the usual limitations of available genetic diversity encountered in conventional breeding.[20] With these methods, scientists are not limited by species boundaries in accessing genetic resources, and can eliminate undesirable traits that might be linked in conventional breeding (Fuchs and Gonslaves, 1997). While these techniques have been applied to soybeans as well as other crop species in laboratory research, there are as yet no commercially available soybean varieties with genetically engineered disease resistance (ISAAA, 2018).

Phytophthora sojae is a highly genetically diverse species. Plant pathologists have identified at least 55 distinct races, and other strains that do not fit into these categories have been reported. Repeated surveys in several US states have indicated that this diversity has been increasing over time, at least partly driven by the deployment of resistance genes (Malvick and Grunden, 2004). For some time, specific resistance genes have been a primary method of *P. sojae* disease management. Recent studies indicate, however, that soybean cultivars with high levels of partial resistance fare better when confronted with novel pathogen strains. *Pythium* spp. also show a great deal of genetic diversity, and many species, as noted earlier, are pathogenic to both maize and soybeans. Robust genetic resistance to *Pythium* spp. infection has been limited. However, plant scientists have identified a few cultivars with varying degrees of resistance. For example, a cultivar known as Archer has shown resistance to a broad range of *Pythium* spp. in field trials (Matthiesen *et al.*, 2016). Research on resistance to *Pythium* spp. is continuing.

Perceived Effectiveness of Disease Control Practices

The preceding discussion describes the range of options available in the disease control toolkit that is available to soybean farmers. The apparent effectiveness of any of these disease control measures is subject to considerable year-to-year and field-to-field variation, making any assessment of their value inherently difficult. Annual variation in weather and other environmental factors means that conditions that would support a significant seedling disease outbreak only occur infrequently. A farmer might see little disease damage in a year when no disease control was attempted and when conditions turned out to be unfavorable for disease development. In contrast, a soybean farmer might see significant damage in a year when a great deal of effort to control disease was expended; the loss may have been much greater in the absence of the effort. Likewise, the very same disease control practices may produce different yield loss outcomes in different fields, based on variations in soil type, pathogen population, and other factors. This sort of stochasticity makes it difficult for farmers to discern and understand the relationship between disease control practices and outcomes, giving rise to model uncertainty.

Information on environmental conditions, the incidence and severity of disease, available and effective control practices, and field conditions can also vary in overall quality from one farmer to another. Not all farmers are equally knowledgeable or skilled, and they do not all have access to the same quality of information (e.g. consultants, soil tests, extension services). Differential information quality can lead to measurement errors and further cloud farmers' understanding between the use of disease control practices and outcomes.

So how do soybean farmers understand the relative effectiveness of all alternative disease control practices and which practices do they use on their farms? In order to answer these questions, we once again turn to our farmer and CCA surveys. In the absence of objective measures of relative effectiveness for the various disease control practices, we treat the CCA survey responses as a baseline against which we can compare farmers' responses.

Soybean farmer and CCA assessments of the effectiveness of seedling disease management methods are shown in Table 6.1. Farmers considered a number of management methods to be effective. Among these agronomic practices, 75–80% of all soybean farmers regarded crop rotation as effective in controlling seedling disease. In contrast, less than 33% of the farmers consider planting later in the season, tilling the land, or adjusting the plant population to be effective control practices. Approximately 70% of soybean farmers rated fungicide seed treatments as effective against seedling disease, but only a third of them expected the use of in-furrow fungicides to be effective. Genetic resistance was also seen as an effective control measure against seedling disease by more than 70% of soybean farmers. Most of these expectations seemed to be in line with those offered by the CCAs, except that they seemed to underestimate the effectiveness of planting date and tillage, especially in the 2013 survey. The CCAs were also more positive than the farmers about the effectiveness of seeds treated with fungicides, although both groups considered this an effective disease control measure.

Farmers' perceptions about the effectiveness of PRR control practices, detailed in Table 6.2, largely paralleled those regarding seedling disease. Among agronomic methods, the only one seen as effective against root rots by a majority (62–65%) of soybean farmers was once again crop rotation. Less than 30% of soybean farmers expected tilling the soil or adjusting the plant population to be effective in controlling PRR. More than 60% of farmers, however, expect that improving soil drainage, a more permanent modification to production fields, to be of comparable effectiveness to crop rotation. Only 40% of farmers expected the use of in-furrow chemical fungicides to be effective in disease control, while some 70% expected genetic resistance to be effective against mid-season rots. The expectations of the CCAs were similar to those of farmers except in the case of fungicide use, where a majority of CCAs considers them effective.

Table 6.1. Survey question on the effectiveness of various practices in managing seedling disease (percentage responding "effective" or "very effective"). (Data from EMAC CCA surveys in 2012 and 2015 and farmer surveys in 2013 and 2016.)

	CCAs (%)		Farmers (%)	
Practice	2012	2015	2013	2016
Till the field	46	45	28	39
Plant later in the season	54	55	19	45
Rotate to another crop	78	82	75	80
Adjust plant population	29	29	28	33
Use fungicides	39	40	33	38
Treat seed with a fungicide	90	90	64	73
Plant a resistant variety	75	80	66	77

Table 6.2. Survey question on the effectiveness in managing PRR (percentage responding "effective" or "very effective"). (Data from EMAC CCA surveys in 2012 and 2015 and farmer surveys in 2013 and 2016.)

	CCAs (%)		Farmers (%)	
	2012	2015	2013	2016
Till the field	39	31	26	33
Improve soil drainage	NA	NA	62	66
Rotate to another crop	63	61	60	69
Adjust plant population	21	18	17	22
Use fungicides	67	68	40	44
Plant resistant varieties	71	76	69	72

NA, Option not available in CCA surveys.

Overall, crop rotation seems to be the only agronomic practice that is expected by a large majority of US soybean farmers to be an effective control for seedling disease and PRR. Planting density, planting date, and tillage practices were perceived as effective disease control methods only by a minority of farmers. Preventive disease control methods, such as the use of genetic resistance in the seed and seed treatments, were expected to be effective by the majority of soybean farmers. Responsive control methods, like the use of fungicide for PRR, were expected to be effective by only a minority of farmers. The subjective expectations of farmers were also consistent with those of the CCAs, and we therefore conclude that, as a group, soybean farmers form effective expectations about the relative effectiveness of alternative disease control methods in their fields.

Use of Disease Control Practices

In light of these expectations, it is important to know which disease control practices farmers actually use to control seedling disease and mid-season root rots on their farms. Once again, we used farmer and CCA surveys to discover such practices.

Agronomic practices

As Table 6.3 shows, soybean farmers use conservation tillage practices quite extensively on their farms. Specifically, they use no-till practices on more than 40% of their soybean acres and minimum till on another 38%. Agronomic practices such as soil tillage may offer some benefits in disease control but forego benefits in other areas. Use of conservation tillage may lower production costs due to fewer passes of the machinery over the field, improve soil health, and improve yields. Farmers consider such opportunity costs in their choices. In fact, we found that US soybean farmers expected the economic benefits of conservation tillage to be significant. Specifically, they expected no-till and reduced-till practices to add approximately three bushels/acre and two bushels/acre to their yields, respectively. Furthermore, farmers expected that no-till and reduced-till practices would decrease input costs by $27/acre and $14/acre, respectively. Hence, it is possible that low farmer ratings of soil

Table 6.3. Farmer survey: tillage practices and benefits. (Data from EMAC farmer surveys in 2013 and 2016.)

Practice	Share of acres (%)		Increased yield (bushels/acre)		Reduced cost ($/acre)
	2013	2016	2013	2016	2013
No-till	49	43	3	2.8	27
Reduced till	37	38	2	2.7	14
Conventional till	14	19	–	–	–

tillage for disease control method are based on their perceived desirability, rather than effectiveness per se, because of the large perceived gains in productivity and cost-efficiency of conservation tillage. This is an important finding, as it suggests that soybean farmers will not regard tillage practices as a relevant method for the control of seedling disease and seed rots under most circumstances. Instead, conservation tillage should be understood as a farming practice that will persist despite its potential contribution to the incidence and severity of such disease.

Soybean farmers also try to plant their soybean crop early in the season; the planting date distribution illustrated in Fig. 6.1 shows that most planting takes place in early May. The ability to plant earlier in the season is perceived by farmers as carrying a significant yield advantage, which they estimate to be roughly five bushels/acre (Table 6.4). This perceived yield gain varies significantly by geography, however. As Fig. 6.2 shows, farmers in southern states expect a greater benefit, as a rule, while the expected yield gain from early planting was lower in most areas where seedling disease and mid-season root rots are frequent.

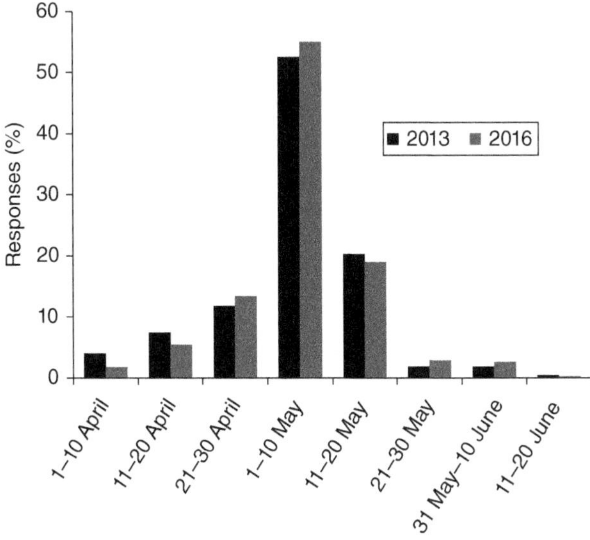

Fig. 6.1. Farmer-reported planting dates. (Data from EMAC farmer survey in 2016.)

Table 6.4. Farmer survey: planting date and early planting benefit. (Data from EMAC farmer surveys in 2013 and 2016.)

	2013	2016
Average planting date	5 May	2 May
Yield benefit (bushels/acre)	5	4.9

Nevertheless, as most soybean farmers expect large potential yield gains from early planting, it is unlikely that they regard the planting date as a relevant method for the control of seedling disease. In fact, earlier planting should be understood as a farming practice that will be continually pursued by farmers despite its potential contribution to the incidence and severity of seed rots and seedling disease. Farmers with significant disease presence on their farms, however, might still think of disease when they decide on planting dates. When CCAs were asked whether farmers would plant their soybeans earlier if seedling diseases were eradicated, a significant share expected that a large number of farmers would do so, given the opportunity (Table 6.5). This suggests that disease considerations may play a role in the decision to plant earlier in the season, effectively limiting adoption for some soybean farmers.

Chemical controls and genetic resistance

Initial survey questions indicated that farmers expect chemical fungicide seed treatments and genetic disease resistance to be among the more effective

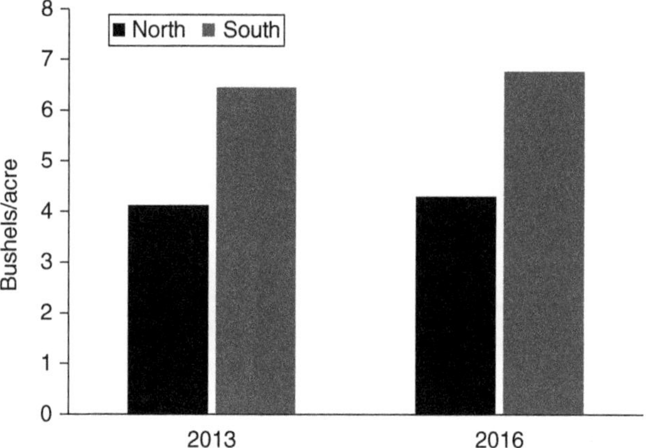

Fig. 6.2. Expected benefit from early planting. (Data from EMAC farmer survey in 2013 and 2016.)

Table 6.5. CCA survey: "If seedling diseases were eradicated in your area, what proportion of growers do you think would plant their soybeans earlier?" (Data from EMAC CCA surveys in 2012 and 2015.)

	2012 (%)	2015 (%)
None	7	11
<25%	38	46
25–50%	30	26
50–75%	15	10
>75%	8	5

disease control methods available to them. Their reported seed purchases suggested that their actions are largely consistent with their subjective expectations. As shown in Table 6.6, 75% of soybean farmers reported purchasing fungicide seed treatments on at least a portion of their soybean seeds, while 50–60% bought seeds with resistance to seedling disease or PRR. The numbers of farmers that bought fungicide seed treatments and genetic resistance in the seed were similar to the number of farmers who expected these technologies to be effective against the diseases (Table 6.1). As only part of the seed that farmers buy is treated with fungicides or includes genetic disease resistance traits, however, the acreage shares of such disease control methods are somewhat lower. Almost 60% of the soybean seed bought by the farmers in our sample included fungicide seed treatments. Estimated shares by the CCAs are shown in Table 6.7 and are comparable to those derived from the farmer survey. Overall, these figures indicate broad adoption of seed treatments for disease control and an overall willingness by three out of four soybean farmers to pay roughly $7.10/acre for a seedling disease management practice with some expected efficacy.

The share of seeds purchased that contain genetic resistance to seedling disease and mid-season rot was lower than that of seeds treated with fungicides, roughly 29–37% according to our farmer surveys. The CCAs estimated such shares to be even lower, 21–26%, as indicated in Table 6.7. These figures suggest that adoption of genetic resistance trails behind that of seed treatments. Nevertheless, resistant cultivars are considered effective for disease control and are broadly used by a majority of soybean farmers on a significant share of their acres. Figure 6.3 also illustrates that genetic resistance and seed

Table 6.6. Farmer survey: purchases of seed with disease control technologies. (Data from EMAC farmer surveys in 2013 and 2016.)

	Farmers who report purchasing (%)		Seed purchased (%)	
Technology	2013	2016	2013	2016
Fungicide treatment	76	75	57	59
Insecticide/inoculant treatment	72	79	54	66
Seedling disease resistance	55	47	35	29
PRR resistance	62	58	37	36

Table 6.7. CCA survey: percentage of seed purchased with disease control technologies. (From EMAC CCA surveys in 2012 and 2015.)

Technology	2012 (%)	2015 (%)
Fungicide treatment	55	60
Insecticide/inoculant treatment	52	57
Seedling disease resistance	21	21
PRR resistance	26	23

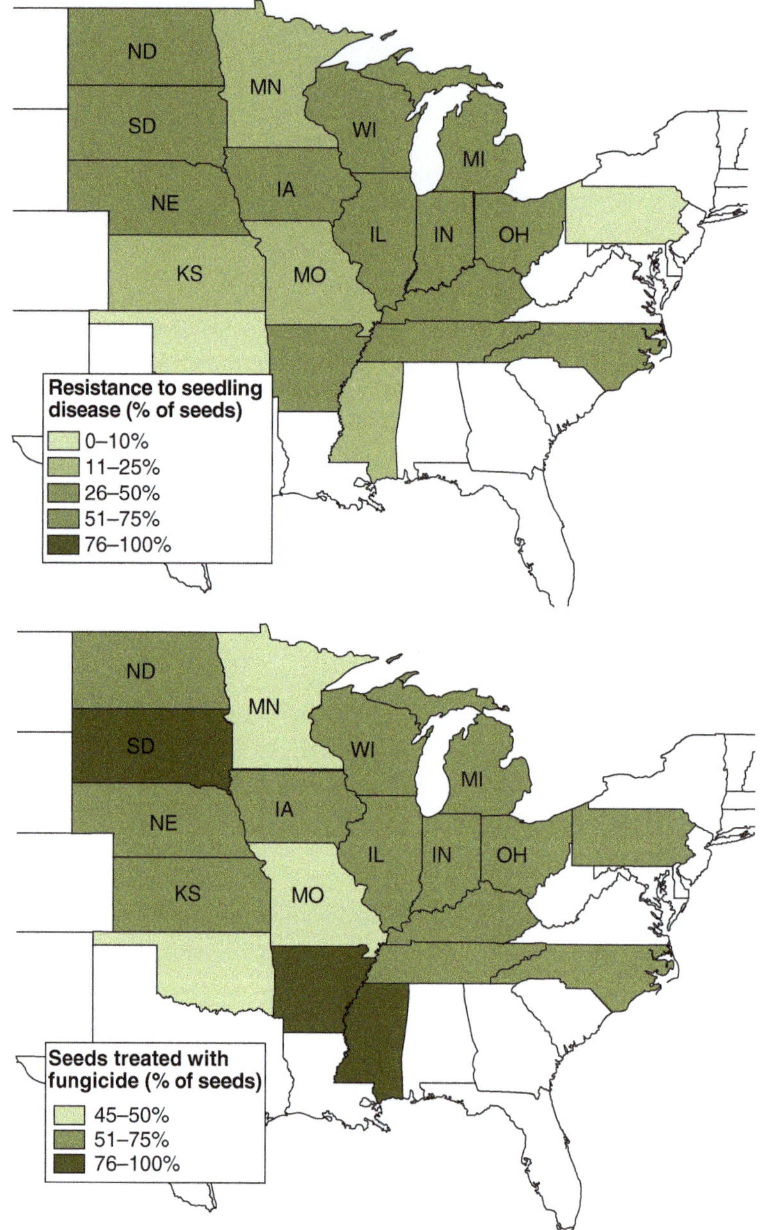

Fig. 6.3. Adoption of seed treatments and genetic resistance. (Data from EMAC farmer survey in 2016.)

treatments are not uniformly adopted across all states. In fact, their patterns of adoption are similarly driven by farmer expectations of effectiveness and economic value as they tend to be greater where the incidence and severity of disease is greater.

Disease incidence and use of control practices

It is interesting to note that while about a third of soybean farmers experienced an occurrence of seedling disease and mid-season seed rots on their farms, twice as many use seeds that are treated with fungicides or include genetic resistance traits. It is therefore clear that some soybean farmers may use these preventive technologies as a precaution, a sort of insurance (Stern, 1973). Alternatively, it may be that because some farmers routinely use preventive disease control practices, they do not actually experience disease on their farms; the direction of causality in this case is unclear. To further evaluate such behavior, we split the producer samples into two groups: one of farmers who reported experiencing disease on their farm and the other of farmers with no such presence.

Responses to both the 2013 and 2016 farmer surveys showed a positive effect of disease history on the farmers' decisions to use the two preventive disease control technologies. For instance, as Fig. 6.4 illustrates, in 2016 some 85% of all soybean farmers who experienced seedling disease on their farm reported purchasing seeds with fungicide treatments, nearly 70% purchased seeds that contained *Rps* resistance, and more than 50% purchased seeds that contained both technologies. A smaller but still significant share of soybean farmers who did not report disease also adopted these technologies. Overall, 70% of farmers with no recent occurrence of disease on their farm purchased treated seeds, over 50% purchased resistant seeds, and over 40% purchased seeds with both traits. Thus, the pattern of broad adoption of preventive measures noted above is not restricted to farmers who are managing ongoing disease problems on their farms. There is another group of farmers who believe that prevention is a prudent course of action. This group may include farmers who see their farms as being at some risk of disease based on environmental conditions, disease incidence in their area, or other factors. In any event, such broad

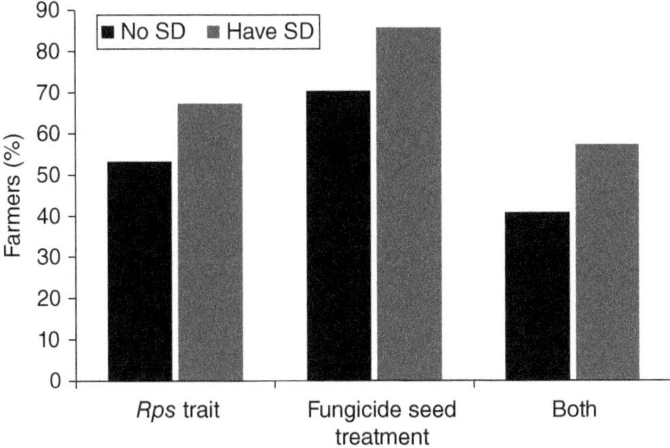

Fig. 6.4. Percentage of farmers purchasing seed with the *Rps* trait, fungicide seed treatment, or both, on farms that have or have not experienced seedling disease (SD). (Data from EMAC 2016 farmer survey.)

adoption of preventive technologies may well be playing an important role in both suppressing disease incidence and severity at the farm level and possibly what is understood nationally.

Insurance, Warranties, and Information

Risk mitigation and disease control

Farmers' choices among the various disease control practices described here may be affected not only by their expected disease incidence and severity on their farm or the expected effectiveness of these practices in controlling disease, but also by risk mitigation practices. Insurance and warranty plans available to farmers can condition potential economic damages and risks, and change farmers' behavior. Using detailed data from our 2013 farmer survey, we found, for instance, that 89% of soybean farmers with stand establishment problems from seedling disease had taken advantage of seed company warranties for replanting, and a further 50% had filed crop insurance claims[21] (Table 6.8). These two insurance mechanisms were perceived by soybean farmers to cover a total of 46% of the associated economic loss. For root rot, 12% had used insurance for losses, which covered 41% of the loss.

From these results, it is apparent that, while farmers rely on insurance to mitigate losses from seedling diseases and mid-season root rots, they do not expect insurance and warranties to cover the entire risk. Nevertheless, by limiting the overall economic risk exposure of soybean farmers from seedling disease and mid-season root rot infestations, insurance and warranties may significantly influence the overall adoption of disease control practices in any given year.

Farmers' expectations and information sources

Throughout this book, we have emphasized the importance of farmers' expectations for effective disease management in the face of extreme uncertainty.

Table 6.8. Farmer survey: use of crop insurance and warranties to mitigate losses from seedling disease. (From EMAC farmer survey in 2013.)

Question	Response (%)
Have you used seed company warranties when you had to replant? (% yes)	89
What share of replant costs do the warranties cover?	54
When stand problems occurred, did you try to recover loss through crop insurance? (% yes)	50
What share of loss was covered by crop insurance?	41
What share of loss was covered by all insurance warranties?	46
Insurance offers "peace of mind" (% agree or strongly agree)	31

We have found that the subjective expectations formed by US soybean farmers about the incidence and severity of disease, potential agronomic and economic losses, and the relevance of alternative control methods are generally effective and consistent with those of experts and with aggregate data. This is an important result because it suggests that soybean farmers can sort through significant observation, model, and process uncertainties in their decision-making process. Importantly, it also indicates that soybean farmers, as a group, are in a position to effectively manage soybean diseases based on the sort of economic considerations described in the damage abatement framework outlined in Chapter 4 (this volume).

The role of reliable and rich information in assisting US soybean farmers in forming effective subjective expectations cannot be overemphasized. In our 2013 survey, we asked soybean farmers about the information sources they use in disease management and their relative importance. What we found was a complex and pluralistic information network that combines expert opinion from both the public and private sector, their own experience, the experience of other farmers, and secondary information sources. Table 6.9 reports the percentage of farmers who reported depending on any one of 18 separate sources of information for the choice of seeds and the number of farmers who indicated that any one of these sources was the most important. The results show that, while seed dealers (in most cases, farmers selling seeds), seed company representatives, and agronomists are significant sources of information for soybean

Table 6.9. Sources of seed-related information for producers. (Data from EMAC 2013 farmer surveys.)

	Percentage who relied on this method	Count of "most important" (n)
Side-by-side on-farm comparisons	85.8	68
A seed company sales representative	83.6	40
A seed dealer	81.6	148
Seed plots of a particular seed company	70.2	10
Seed plots of a seed dealer	68.0	19
Another farmer	67.4	18
Another farmer's field	66.2	9
A seed company agronomist	60.2	21
University seed plots	58.8	58
State variety trials	58.0	25
A seed company catalog	47.4	5
A university agronomist	45.8	15
Other, third-party variety trials (crop adviser/ banks/farm managers)	34.6	10
An article in a magazine/newspaper	34.0	4
The website of a seed company/university/seed dealer	32.6	0
Farm shows where they talked about seed	27.6	1
A paid crop adviser	24.0	22
An e-mail from a seed company	10.8	0

farmers, side-by-side experiments on their farm, seed plots in universities, and other farmers' fields were also important sources. Various secondary sources of data, from magazine publications to websites, provided important additional references. These results suggest that, in the presence of extreme uncertainty, such as that faced by soybean farmers in the context of disease control, access to high-quality, timely, and decision-relevant information from multiple sources can provide a valuable countermeasure.

Summary

Farmers have three general categories of disease control methods available to them: agronomic practices, chemical controls, and genetic resistance in the seed. Choosing among them often has implications beyond simple disease management goals, impacting the production process and farm profitability. Current agronomic practices, such as conservation tillage and early planting, have cost-efficiencies and yield benefits that appear to far outweigh, for most farmers, whatever negative impact they might have on disease incidence and severity. Pre-emptive disease control practices, such as the use of genetic resistance and fungicide seed treatments, are more broadly used and are expected to be effective by most farmers, including many with no recent disease incidence experience. Preventive practices thus serve a risk management function, along with disease suppression. Responsive control practices are used only in limited ways. Risk management practices, such as crop insurance and replant warranties, can condition farmer disease management decision making and involve considerations that go beyond disease pathogen control. Despite facing extreme uncertainty, soybean farmers, as a group, appear capable of forming reliable expectations about disease incidence, severity, and yield loss, as well as about the means to control the disease. For their management decisions, farmers rely on many different information sources, both private and public. Access to high-quality information from multiple, diverse sources can help offset extreme uncertainty and assist farmers in forming accurate and reliable expectations for effective disease management.

Notes

[20] See Chapter 9 (this volume) for additional details on gene-editing technologies.

[21] The USDA underwrites and administers crop insurance policies that are sold and serviced by private agents. The policies insure farmers against yield losses from natural causes such as extreme weather, insects, and disease.

7   Costs, Profits, and Farm Demand for Disease Control

In its most basic form, economic decision making consists of gathering information and maximizing benefits relative to costs over a set of alternative choices. In the context of disease control, farmers gather information on the potential yield losses from pathogens, the various methods that control them, and their costs, and choose the ones that maximize their revenues relative to their costs. In this chapter, we consider how soybean farmers choose disease control methods for their farms and how their collective choices add up to form the aggregate demand for disease control inputs in soybean production.

Soybean farmers begin by forming expectations about the potential yield loss from pathogens and the effectiveness of control practices, as discussed in the previous two chapters. To make an economic decision as to which control practice to use and to what extent, soybean farmers must also form expectations about their costs and revenue potential from yield savings. Forming cost expectations across disease control practices requires different amounts of information and effort. For some disease control practices, there are opportunity costs, which can differ in size across farming operations. For instance, some farmers expect higher soybean yields from early planting, and for these farmers, planting later in the season for better disease control would come at a loss of yield and revenue opportunity – an implied cost. Soybean farmers form expectations for such opportunity costs based on their own cropping practices and yields. Other disease control practices involve inputs traded in the market with transparent prices in each transaction, such as fungicides or seed treatments. Forming cost expectations for such disease control practices is therefore more straightforward for farmers. Finally, there are disease control practices that involve inputs traded in the market but at prices that are not transparent. For example, disease resistance in the seed is sold as a bundle with other genetic traits (e.g. drought tolerance, herbicide tolerance, etc.) at a single overall price per bag of seed. For such control practices, forming cost expectations might be somewhat more demanding.

To form expectations about the revenue potential of disease control practices, soybean farmers must first form expectations about the price they will receive for their crop. Soybean prices change from day to day and vary from one location to another (Helmberger and Chavas, 1996). Nevertheless, soybean price information is readily available in any given day at local and national levels, and farmers can form expectations of their sale price to guide the economic decisions on their farm, including disease management (e.g. Trading Economics, 2018). Overall, soybean farmers must deal with varying degrees of complexity in forming expectations on the costs and revenues of disease controls and they must regularly update these expectations in order to maximize profits, as discussed in Chapter 4 (this volume).

In this chapter, we examine how soybean farmers choose disease control practices that maximize their profits, in practice. For our more detailed analysis in this chapter, we once again focus on seedling disease and mid-season rots in US soybean production. Farmers' expectations of opportunity costs for agronomic practices that can control these specific diseases were discussed in Chapter 6 (this volume); cost expectations of fungicides and seed treatments are assumed to be consistent with observed market prices. Here, we begin by examining the cost of genetic resistance against seedling disease and PRR, which is not directly priced in the market but can be estimated. We then discuss in some detail how farmers use their subjective expectations about the incidence and severity of disease, as well as about the relative effectiveness, cost, and revenue potential of control practices in order to maximize farm profits.

Seed Traits and Implicit Prices for Disease Resistance

Farmers purchase soybean seed as a bundle of traits, including: yield potential, relative maturity, plant height, and other agronomic characteristics; herbicide tolerance; specific or partial resistance to various diseases; and others. Soybean varieties therefore differ from one another in the traits they contain and in their performance, and, as such, they command different prices in the market. Here, we examine how much more expensive soybean seeds may be when they include resistance to seedling disease and PRR.

In the hedonic pricing method pioneered by Lancaster (1966), buyers derive satisfaction, or utility, not from a particular good per se, but rather from the specific attributes of the good that satisfy the buyers' needs. The market price of a good is thus the result of an equilibrium between buyers' subjective valuations of, and thus willingness to pay for, each relevant attribute and sellers' willingness to accept a particular price for each attribute. None of the attribute prices is stated explicitly in the market, however; the price of the bundled good is made up of the implicit prices of all the relevant attributes, or:

$$P = \sum_{i=1}^{n} p_i \tag{7.1}$$

where P is the market price of the good containing n relevant attributes and $\boldsymbol{p}$ is the vector of implicit prices of the attributes, i. In empirical studies, the relationship is usually operationalized as:

$$P = \sum_{i=1}^{n} \beta_i X_i \tag{7.2}$$

where $\boldsymbol{X}$ is the vector of variables indicating the presence of the relevant attributes and $\boldsymbol{\beta}$ is the vector of weights indicating the importance of each attribute in determining the final price of the good (Espinosa and Goodwin, 1991; Dalton, 2004).

This hedonic pricing approach has been applied to agricultural input markets (e.g. Beach and Carlson, 1993; Barkley and Porter, 1996; Kristofersson and Rickertsen, 2004) and can be applied in the soybean seed market as well. In this context, when farmers purchase soybean seed, they are looking for the bundle of agronomic, production, and other traits that they expect will fit best with their production systems and will maximize their farm profits. Genetic disease resistance is just one trait among many, each with its own implicit price. By analyzing the market prices of a wide variety of soybean cultivars with different traits, we can estimate the implicit price of genetic resistance to seeding disease and PRR that farmers pay in the US market.

As before, our analysis focuses on seedling disease and mid-season rots to gain detailed insight. In this context, we first obtained price information for soybean seed varieties sold in the US from a commercial database.[22] The database includes thousands of soybean seed varieties sold in the USA, their average prices, and estimated area planted for each variety. We then collected trait information by variety in order to link varietal attributes to seed prices. Information on seed traits was pieced together from a number of alternative sources, including state seed variety trials (e.g. Ames *et al.*, 2016), company seed catalogs, and plant variety protection certificates (PVPO, 2017). The specific traits reported varied across sources, so we included only the most common traits in our analysis: herbicide tolerance (e.g. Roundup Ready®) and disease resistance (e.g. *Rps* and SCN genes). In addition, we included a variable identifying seed from the top seed brands, in terms of market share, to account for the possibility that these might be viewed as elite seed lines and thus command a price premium.

Various functional forms can be used to specify hedonic pricing models. For this study, we chose the semi-logarithmic form, which allows implied price premiums for different traits to be calculated directly from the β parameters using the formula $\boldsymbol{p}_i = (\exp(\boldsymbol{\beta}_i) - 1) \times 100$ (Wooldridge, 2009, pp. 232–233). Other functional forms yielded similar results. The specific function used here is:

$$\ln Price_i = \beta_0 + \beta_1 Roundup\ Ready_i + \beta_2 Roundup\ Ready\ 2\,Yield_i$$
$$+ \beta_3 LibertyLink_i + \beta_4 STS_i + \beta_5 SCN_i + \beta_6 Rps_i + \beta_7 Brand_i + \varepsilon_i \tag{7.3}$$

where ε_i is the error term for the regression equation.

The results of this model are shown in Table 7.1. The estimated premium for STS® (sulfonylurea-tolerant soybean), the oldest available herbicide tolerance trait, is small and not statistically significant. While our model calculated

Table 7.1. Seed trait pricing model. (Data from industry sources and author models.)

Variable	Parameter estimate	Standard error	P value	Price premium (%)
Intercept	3.267***	0.051	<0.0001	
Roundup Ready®	0.212***	0.036	<0.0001	23.67
Roundup Ready 2 Yield®	0.367***	0.038	<0.0001	44.33
LibertyLink®	0.332***	0.106	0.0018	39.39
STS®	0.034	0.076	0.651	3.49
SCN	0.090**	0.038	0.0181	9.37
Rps	0.081***	0.026	0.0022	8.43
Brand equity	0.152	0.118	0.2013	16.37

***, Statistically significant at the 1% level; **, statistically significant at the 5% level.

a sizable 16% price premium for the top brands, it was not statistically significant. All other estimated price premiums were statistically significant at conventional levels. The model indicates that soybean farmers pay an average 8.4% premium for seed varieties that contain *Rps* resistance traits, which amounts to around $4.10 per bag at current prices. This compares to a roughly $4.60 per bag implicit premium for SCN resistance, and higher premiums of $11.50 per bag for the original Roundup Ready® trait, $21.60 for Roundup Ready 2 Yield®, and $19.20 for LibertyLink®. Such implicit prices must therefore be paid by farmers who are interested in the individual traits. We expect soybean farmers to form expectations about the implied costs of the traits they purchase.

After forming expectations about the incidence and severity of disease, the relative effectiveness of the various control practices, and their associated costs and revenue potential, soybean farmers can maximize expected profits as described by Eqn 4.4 in the standard damage abatement model. Indeed, the model clarifies that farmers can ascertain whether the expected yield damage and revenue loss from disease justifies the cost of control and, if so, apply control measures up to the point where the incremental cost is just equal to the value of the yield loss prevented.[23]

The Calculus of Profit Maximization

While the principle of maximizing profits under the damage abatement model is straightforward, the implied calculus turns out to be quite complex. To illustrate the mechanics of the farmers' economic decision-making process, we use here a decision tree framework (Hyde *et al.*, 1999, 2003). Decision tree models involve tree-like depictions of alternative choices and outcomes and are used in operations research and other fields to analyze economic decision making. They are instructive as they mimic the decision process and identify the information needed for choosing economically optimal solutions.

A decision tree model of disease control

In the context of disease control, the decision tree model can simulate the farmer's decision process of choosing among different control practices while accounting for their expected effectiveness and costs as well as the different states of the environment and markets that condition them. In order to account for multiple sources of uncertainty, the model incorporates stochasticity through Monte Carlo simulation (Yoe, 2011), where soybean prices, yields, disease pressure, and other relevant variables fluctuate according to empirical distributions. The estimated outcome from each simulated decision iteratively populates a farm budget and returns a calculated pay-off. Optimum disease control strategies can then be identified through direct comparisons of all pay-offs and their relative contributions to farm profits.

In any given year, soybean farmers begin by forming expectations about the incidence and severity of disease in their fields. These expectations are conditional on prior decisions that determine the crop rotation and tillage practices used on the farm, as well as on expected environmental conditions. The farmer also forms expectations about the relative effectiveness of disease control practices and likely market conditions. From these comes a cascade of follow-on decisions necessary to implement a functional production plan that might include disease control. Among these are decisions about crop genetics and disease resistance in the seed, seed treatments, crop insurance, planting date, scouting fields for disease symptoms, fungicide applications, replanting, and others. Each decision point creates a new branch of follow-on decisions. When all of the potential decision paths are laid out, the complexity of the decision is evident; the model yields many thousands of possible outcomes.

For our discussion here, we once again focus on seedling disease and mid-season rots and, for the sake of simplicity, we consider only a small subset of the farmer's decisions by limiting the potential alternatives to a few considerations. Specifically, we consider the farmer's decision whether to use either of the two most common practices for controlling seedling disease and mid-season rots: seed treatment and resistance genes. Figure 7.1 provides a simplified view of the decision tree model structure. The model begins with the producer purchasing soybean seed that either contains *Rps* resistance genes or does not. Next the farmer decides whether to apply seed treatments to that seed. Each of these two decisions determines the level of protection against potential seedling disease and mid-season root rots. However, only after these decisions are made, and the soybeans are planted, is the actual disease pressure realized. For our decision tree model, we specify that, in any given year, soybean farmers face a 24% chance that they will experience seedling disease (2.4 years out of 10) with an associated yield loss of 8.2 bushels/acre, when the disease is present.[24] Of course, this damage may be partially offset by the control decisions made by the farmer at the beginning of the cropping season. If there is no incidence of disease in the farmer's fields, no yield loss is realized. In either case, the farmer also faces a 22.7% chance (2.27 years out of 10) of incurring a 7.9 bushels/acre yield loss to mid-season root rot. Each decision and outcome results in a separate farm budget and determines the decision's effect on farm

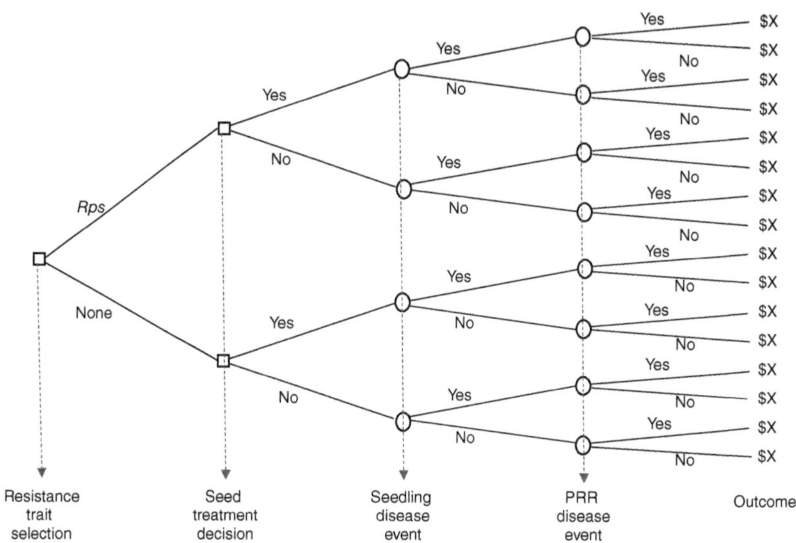

Fig. 7.1. Decision tree model.

profitability. The farm budget inserts soybean prices and yields into the revenue calculation and subtracts the costs of genetic resistance in the seed and seed treatments.

Using this simple decision tree, we consider two scenarios. In the first, the producer only has to manage seedling disease. In the second, both seedling disease and mid-season root rots, specifically PRR, are to be managed.[25] In each of these cases, the farmer can choose from seed treatments and *Rps* resistance traits, and the purchased seed may contain either, both, or neither. The *Rps* genes decrease the yield loss from seedling disease by an average of 24% and from PRR by 51%.[26] Seed treatments decrease the yield loss from seedling disease by an average of 59% and are ineffective against PRR.[27] The cost of employing *Rps* genes is $4.10/acre, as calculated from Table 7.1, while seed treatments cost $7.10/acre, as identified from our farmer surveys. A farmer employing both measures would thus incur a cost of $11.20/acre.

In order to account for the inherent uncertainty, soybean price (USDA, 2017c), soybean yields (USDA, 2018), yield losses from the pathogens, and yield benefits from the two disease control practices varied according to empirical distributions. Their values were determined from independent draws from these distributions over 1000 iterations. Each iteration in the Monte Carlo simulation can be thought of as a unique set of disease and environmental conditions that might pertain to a given field. Each unique iteration generates a different set of pay-off values for the control practices considered, and each has the potential to be the optimum choice, based on the relative pay-off amounts. The pay-offs reported below are averages across all iterations, while the optimum solution percentages can be interpreted as the proportion of planted acres on which a particular set of disease management measures is the profit-maximizing choice.[28]

Baseline analysis

The baseline analysis examines the optimum solutions for the two above scenarios, assuming in all cases that farmers are risk neutral, i.e. they make decisions solely on the basis of the expected pay-off while the variance in the pay-off does not concern them. The baseline results are shown in Table 7.2. The first two columns show the average estimated monetary pay-offs ($/acre) for the few disease control practices examined, compared with taking no action to control seedling disease or mid-season rots. In a situation where the farmer is only managing seedling disease, seed treatment has the highest pay-off, worth $6.30 after the cost of treatment has been paid. *Rps* resistance genes contribute less to farm profitability with a net pay-off of $1.66 when used alone and $5.05 when combined with seed treatment. *Rps* genes therefore reduce the overall pay-off from seed treatment when the two practices are combined as their incremental mitigation of disease is limited and does not pay for their combined cost. Optimum choices change when farmers must deal with PRR in addition to seedling disease. Here, the combination of seed treatment and *Rps* resistance is the most effective at managing the tandem disease condition, preventing yield loss with a net value of $15.61/acre. When compared with the initial cost of $11.20/acre paid by the farmer, this pay-off value represents a net return on investment (ROI) of 139%, the highest of the options considered.

Despite the simplicity of the decision set in our analysis, the results are instructive and informative. An important insight is that the disease control practice with the highest average net pay-off across all conditions is not necessarily the optimum solution in every case. Differing disease conditions may well call for different practices. When dealing with seedling disease alone, our simulations indicated that taking no action to control the disease is the most economical choice in 17% of iterations. This would correspond to the proportion of acres where the expected loss did not surpass the EIL. In the majority of cases, seed treatment alone is the optimum choice, although adding *Rps* resistance actually provided a greater benefit for 8% of acres. When managing both seedling disease and PRR, the highest pay-off practice, the combination

Table 7.2. Optimal choices among selected disease control practices. (Data from the USDA, industry, EMAC 2016 farmer survey, and author models.)

Disease control practice	Pay-off of disease control ($/acre)		Share of optimal solutions		
	Seedling disease only	Seedling disease + PRR	Seedling disease only	Seedling disease + PRR	Farmers' responses from national survey
No action	–	–	17%	0%	13%
Seed treatment only	$6.30	$6.36	75%	0%	25%
Rps genes only	$1.66	$12.15	0%	31%	12%
Seed treatment + *Rps* genes	$5.05	$15.61	8%	69%	50%

of seed treatment and *Rps* resistance is the best choice for the majority of the acres, while *Rps* resistance alone is optimal for 31% of acres.

The results from the simulations also match up well with our national farmer survey responses, shown in the last column of Table 7.2. Percentages in this column correspond to the practices used by US soybean farmers who indicated in our 2016 national survey that they had experienced seedling disease, mid-season root rot, or both in their fields. Overall, 13% of these farmers reported not using either seed treatments or *Rps* genes, which is similar to the share of optimum solutions yielded by the model, when only seedling disease was a potential problem. Seventy-five percent of farmers reported using seed treatments, with 25% using them exclusively, which appears appropriate considering that this is the dominant simulated strategy when only seedling diseases were possible. Fifty percent of farmers reported using both *Rps* genes and seed treatments, which was the dominant simulated strategy when both seedling disease and PRR were possible. In all cases, these numbers lie in between the percentages calculated for the two disease conditions in the model simulations, and hence simulated optimum solutions and actual practices are generally consistent.

Changing Market and Disease Control Conditions

When farmers' expectations about the incidence and severity of the disease, the effectiveness of control practices, or input costs and output prices change, the expected pay-offs of individual practices can change as well. Using sensitivity analysis, we examine here how differing soybean prices, seedling disease incidence, and seed treatment efficacy impact the farmer's optimum choice in disease control. In each case, we hold the variable of interest constant at each of a set of discrete values as we repeat the analysis. Throughout this process, we once again estimate the pay-offs of the three disease control practices – seed treatment, *Rps* resistance, and the combination of these two – when both seedling disease and mid-season rot are potentially present and, as before, we present these pay-offs as net changes against the option of using no disease control.

The results from this sensitivity analysis clarify that the expected price of soybeans is critical to the farmer's decision to control disease. As Fig. 7.2 illustrates, at very low soybean prices, none of the three control practices is economical and the farmer's optimum choice is to take no action to control the disease. At approximately $6/bushel, use of seed treatment breaks even, while the use of *Rps* genes yields a $4.20/acre pay-off. As the expected price of the crop increases, so does the value of yield lost to disease. If soybean prices exceed $9/bushel, the combined use of seed treatment and *Rps* resistance becomes uniformly the optimum choice for disease control. At $9/bushel, the combined practices give an average ROI of 78%, which only increases with higher soybean prices.

Farmers' expectations of disease incidence are also a determining factor. Farmers combine their expectations about incidence and severity to estimate the expected yield loss to disease in any given year. At low incidence levels,

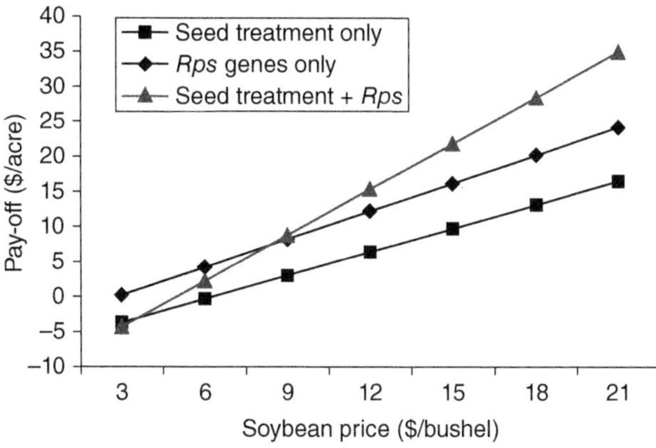

Fig. 7.2. Effect of changing soybean price. (Data from EMAC farmer survey in 2013 and author models.)

expressed as the number of years in 10 years that disease is present, the average expected loss is low and the profit-maximizing choice for soybean farmers is, generally, to take no action to control the disease. Figure 7.3 shows the effect of increasing seedling disease incidence on the economic value of the three disease control practices. At low rates of incidence, *Rps* resistance alone is the most valuable option, giving a return of nearly $7/acre over the alternative of no action and a 62% ROI. This is because PRR incidence is the predominant disease problem. At a seedling disease incidence of 20%, i.e. when the disease is present in 2 years out of 10, the combination of seed treatment and *Rps* resistance becomes the optimum option, with a pay-off of $12.54/acre, a 112% ROI, versus $11.20 from *Rps* resistance alone. At higher incidence rates of seedling disease, the combination of practices maintains its position of primacy and sees even greater ROI.

Changes in the expected effectiveness of disease control practices, which enter the damage abatement model through the kill function $h(N_0, X, A)$, can also change a farmer's optimum choice. Here, we consider the effect of varying efficacy of seed treatment on the value of the three control options. As Fig. 7.4 shows, at a low efficacy level, the use of seed treatment returns a net loss; its efficacy must exceed 30% to make a positive contribution to farm profits. The efficiency of seed treatment must surpass 90% to return a higher pay-off than *Rps* genes used alone. The value of the combination of seed treatment and *Rps* increases as the efficacy of seed treatment improves. When seed treatment efficacy exceeds 45%, the value of the combined method surpasses that of *Rps* resistance used alone. Below this level, *Rps* resistance alone is the optimum method; above it, the combined measure is the optimum.

Finally, the relative cost of each individual disease control practice determines whether soybean farmers will use it in their fields and to what extent. In fact, plotting the changing share of acres for any one disease control practice against its changing unit cost produces the familiar downward-sloping demand curve. Here, we derive the demand curve for one of the control practices, seed treatment. To do so, we vary the cost of seed treatment between $3.50 and

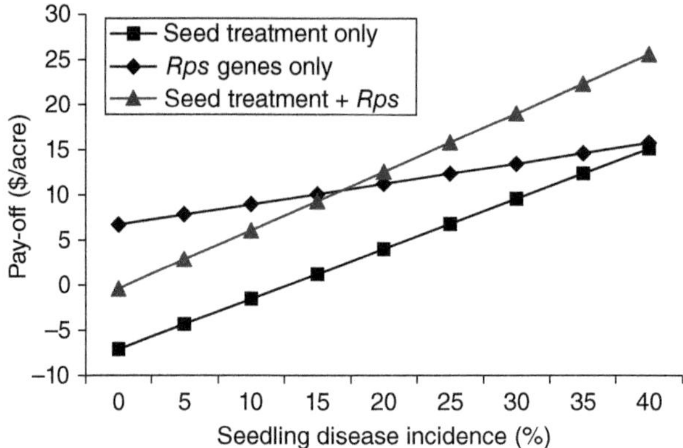

Fig. 7.3. Effect of varying seedling disease incidence. (Data from EMAC farmer survey in 2013 and author models.)

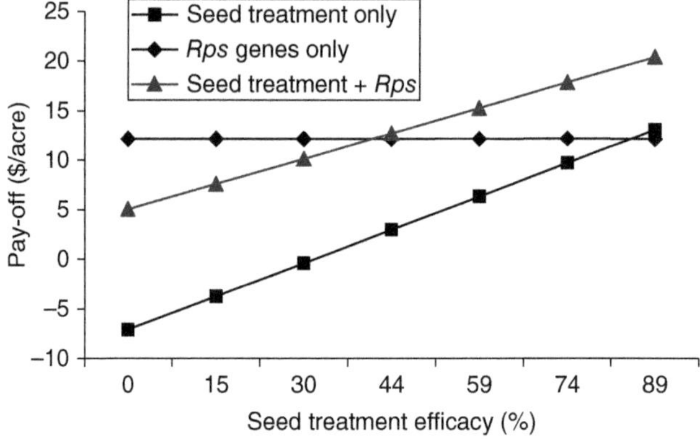

Fig. 7.4. Effect of varying seed treatment efficacy. (Data from EMAC farmer survey in 2013 and author models.)

$17 per bag of seed and for each selected unit cost we solve for the farmer's optimum choice. As before, we estimate the pay-offs of the three disease control practices – seed treatment, *Rps* resistance, and the combination of these two – when both seedling disease and mid-season rot are potentially present. We then plot the projected share of acres that use seed treatment against the unit cost. As Fig. 7.5 illustrates, when the cost of seed treatment declines from a high of $17 per bag to a low of $3.50 per bag, the share of acres that profit-maximizing farmers allocate to seed treatment increases from a low of 10% to a high of 96%. Hence, the demand curve in Fig. 7.5 illustrates the standard inverse relationship between unit costs and disease control practices, and clarifies the expected farmer's response in the face of market swings.

Overall, the sensitivity analysis performed in this section demonstrates that changes in the farmer's environment and market conditions or shifts in the efficacy of control methods can lead to drastically different optimum decisions

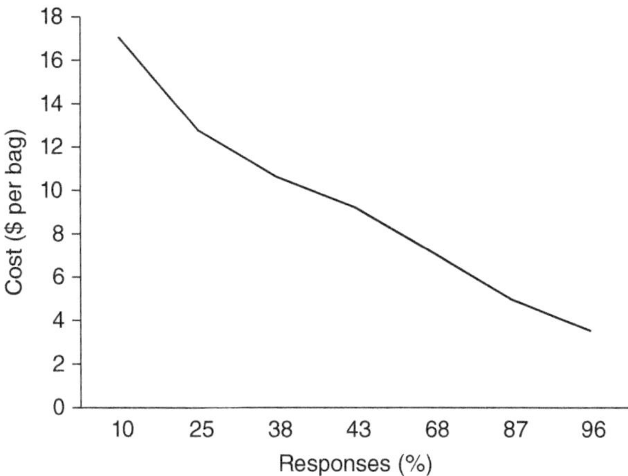

Fig. 7.5. Demand curve for fungicide seed treatment. (Data from EMAC farmer survey in 2013 and author models.)

and pay-offs. As such, farmers' expectations about all of these factors must constantly adapt as new information becomes available.

The Calculus of Risk-averse Farmers

The characteristic of risk neutrality used in the foregoing analysis means that farmers' utility is defined as strictly proportional to the amount of monetary pay-off, and thus their utility maximization may be represented by expected profit maximization. However, farmers, like the rest of us, are often risk-averse. For a risk-averse individual, maximizing expected profits is important, but so is minimizing potential loss. As such, risk aversion changes the farmer's optimization calculus. Instead of analyzing control practices solely on the basis of their expected pay-offs, farmers also consider the certainty of the pay-off in their evaluation. Certainty is generally defined in terms of both the range of possible outcomes, from the highest potential profit to the greatest potential loss, and the relative subjective probabilities assigned to those outcomes. Thus, risk-averse famers may prefer a lower potential profit that is more certain over a higher potential profit that is coupled to a significant potential loss (Kahneman and Tversky, 1979). To examine the impact of risk aversion on the farmer's economic considerations of disease control, we repeat our decision tree analysis, assuming that farmers are no longer risk neutral but rather risk-averse.

With this new calculus, the pay-off amounts of disease control practices are expressed as the monetary equivalents of their utility to the risk-averse farmer. We use a negative exponential utility function here, expressed as $U(W) = -exp(-\rho W)$, where W is the amount of the pay-off and ρ is the coefficient of risk aversion (Hyde *et al.*, 2003). The subjective value of each disease control measure can then be thought of as comprising two parts: the monetary pay-off and the money equivalent of the utility that the farmer derives from the risk-reducing feature of the disease management measure. The greater the risk aversion, the higher the value that is placed on risk reduction. Following Hyde *et al.* (1999), this model specifies relative risk aversion (RRA) to be equal to ρW. We calculate the pay-off for the selected disease control practices under four RRA levels: 0 (the risk-neutral baseline case), 1, 3, and 5, where 1 represents low risk aversion and 5 corresponds well to a highly risk-averse farmer, as demonstrated by Anderson *et al.* (1985).

As shown in Figs 7.6 and 7.7, all disease control options become more valuable as RRA increases. Moreover, the relative values of the different options change as well. In the case where seedling disease alone is managed and for the most risk-averse farmer, the value of seed treatment increases by 2.5-fold relative to the risk-neutral baseline, while the value of *Rps* genes more than quadruples and the value of the combination increases by more than 4.5-fold. This pattern is repeated in the case when both seedling disease and PRR must be managed, but not to the same extent. Here, the value of the combination of

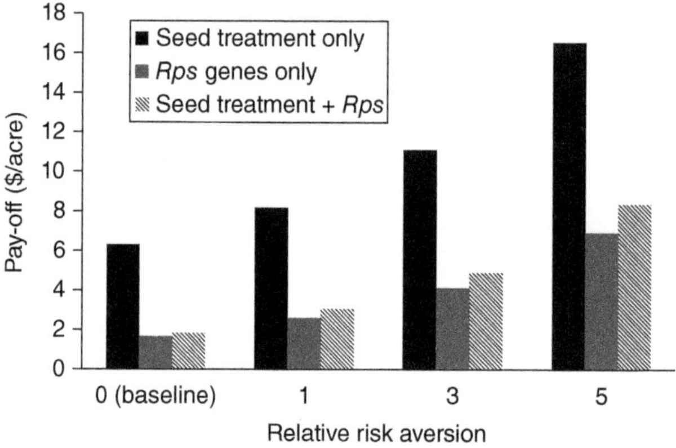

Fig. 7.6. Effect of risk aversion on pay-off values for seedling disease control. (Data from EMAC farmer survey in 2013 and author models.)

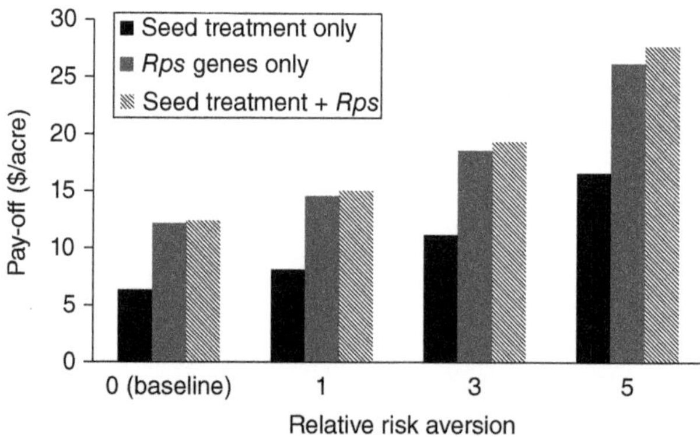

Fig. 7.7. Effect of risk aversion on pay-off values for seedling disease plus PRR. (Data from EMAC farmer survey in 2013 and author models.)

seed treatment and *Rps* resistance more than doubles, while the value of seed treatment alone increases by more than 2.5-fold.

The effects of risk aversion on the share of optimum solutions for the selected control practices are illustrated in Figs 7.8 and 7.9. In all instances, the more risk-reducing practices increase their share as risk aversion increases. When managing seedling disease alone, the combination of seed treatment and *Rps* resistance changes from being the optimum solution on only 8% to 27% of acres as RRA increases. Although seed treatment alone retains its primary role, it consistently decreases its share as RRA increases. Similarly, when farmers are managing seedling disease and mid-season root rot, the combination measure, already the optimum solution in the majority of cases, increases

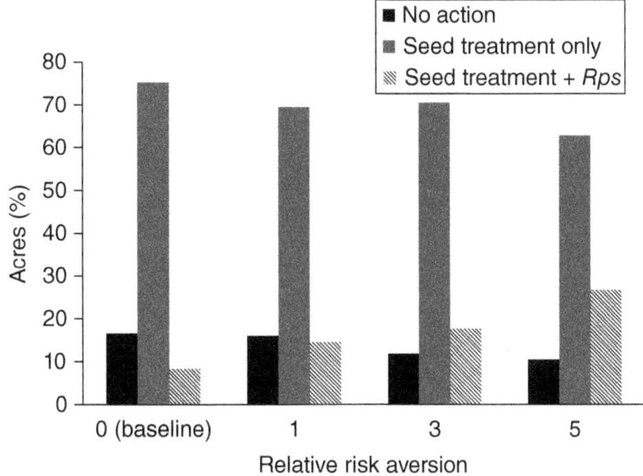

Fig. 7.8. Effect of risk aversion on optimum solutions for seedling disease. (Data from EMAC farmer survey in 2013 and author models.)

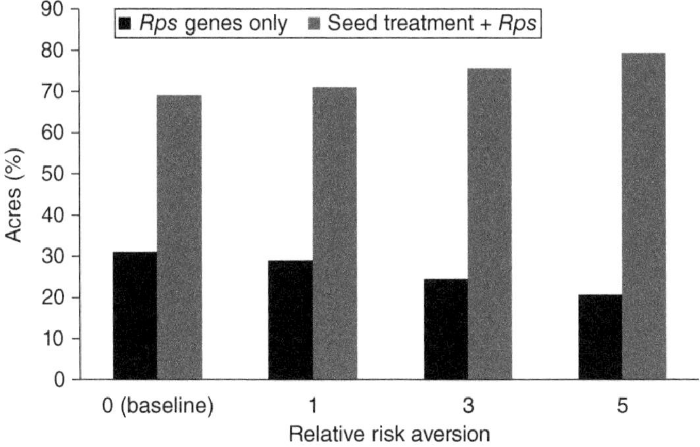

Fig. 7.9. Effect of risk aversion on optimum solutions for seedling disease plus PRR. (Data from EMAC farmer survey in 2013 and author models.)

its share as RRA increases, while *Rps* resistance alone is the optimum solution on fewer and fewer acres.

Farmer Heterogeneity and Aggregate Demand

A recurring theme over the last four chapters has been heterogeneity, both of farms and of farmers. Environmental conditions, such as weather patterns, vary across regions. Local agronomic and disease conditions, such as soil types and pathogen populations, vary across farms and fields. Farmers differ in their knowledge and experience, information, and attitudes toward risk. Soybean prices and input costs vary across geographies and over time. This inherent heterogeneity then leads to different optimum choices for disease control and is behind the familiar downward-sloping aggregate demand curves for disease control inputs observed in the market.

In the process of maximizing their profits or utility under their own unique conditions, farmers implicitly develop individual demand schedules for disease control measures. We can think of the demand schedules of individual farmers as the amounts of a disease control input they buy (e.g. bags of fungicide-treated seed) at different input prices, based on their subjective expectations of the potential benefits. Adding up the individual demand schedules of all soybean farmers would yield a complete market demand curve for a disease control input. Of course, individual demand schedules are not typically observed; the total input quantities purchased by all farmers at specific market prices are observed instead.

Market prices for disease control inputs are set by the input suppliers; farmers have no direct influence and must accept the market price as given. Thus, the farmer's decision is to buy or not buy a certain quantity, based on the farm areas where the expected profit from disease control is greater than, or equal to, the costs of control. At lower market prices, disease control will be profitable on lower-value fields, so the quantity purchased by farmers will be greater. At higher market prices, control will be profitable only on fields with greater losses to disease, so the quantity of input purchased will be smaller. The actual market quantity is therefore determined by the interaction of input suppliers, who decide how much to supply at any given price, and farmers, who decide how much to buy at a given price, based on the economic value of the control they receive.

Market demand for a disease control input is strictly the relationship between the price of the input and the quantity that farmers, as a group, are willing to buy. This representation assumes that all other relevant factors are held constant. If other factors change (e.g. soybean price or the effectiveness of a particular input), the nature of the price–quantity relationship changes. In the context of the demand curve illustrated in Fig. 7.5, the curve may shift to the right or left, indicating a greater or smaller quantity demanded at any given price, and/or the curve may rotate, indicating that the quantity demanded may change by a greater or smaller amount with any given change in price.

Summary

Within the context of disease control, soybean farmers gather information from a variety of sources, including their own direct experience, and form subjective expectations regarding disease incidence and severity, the effectiveness of the various control measures, and market prices. Based on these expectations, they choose the practices that maximize farm profits, or, more broadly, their utility. Even small changes in farmer expectations can shift the optimum choice for disease control. As a result, soybean farmers must regularly update their expectations in order to make effective economic decisions in disease control. Actual disease control practices that soybean farmers use in their fields are found to be consistent with the practices that maximize profits in stylized economic models. Hence, soybean farmers appear to make effective economic decisions in disease control in spite of incomplete information, uncertainty, and intense computational demands in their profit-maximizing calculus. Farmers and farms are, however, heterogeneous in terms of environmental conditions, pathogen populations, farming systems, risk attitudes, and other factors leading to different optimal choices in disease control. These differences underpin the movements in the aggregate demand for disease control inputs that occur when input prices change. In sum, each farmer's rational economic decision making, with the goal of profit maximization, is the foundation of individual and aggregate market demand for disease control practices and inputs.

Notes

[22] The price and varietal data for our analysis was obtained from GfK, a market research company.

[23] Economists refer to this optimality condition as one where the marginal factor cost of an input is equal to its marginal value product. This condition is illustrated for disease control inputs in Fig. 4.1.

[24] These values are derived from the expected incidence and severity reported from the 2016 farm survey discussed in Chapter 5 (this volume).

[25] Note that the farmers in the decision tree analyzed here correspond to the farmer group with some incidence of seedling disease and/or mid-season rots on their farms, roughly a third of all soybean farmers in the USA (see Chapter 5, this volume).

[26] It is assumed here that all soybean varieties with *Rps* genes also include partial resistance, which is the source of the 24% reduction in the yield loss from seedling disease.

[27] Assumptions about efficacy are based on Dorrance *et al.* (2003b, 2009) and Radmer *et al.* (2017).

[28] Farm profits involve whole-farm considerations where the use of all inputs (capital, labor, land, pest control, etc.) is optimized. Here, we consider only a very narrow pay-off concept, which involves the value of the yield loss avoided minus the cost of disease control used. While this simplification is attractive for illustration purposes, we note that farmers face more comprehensive and complex economic decisions when they consider disease control.

Farm Demand for Innovation in Disease Control

To improve disease control, soybean farmers must use better practices. Improved practices can increase yields, reduce production costs, or bring about both. Farmers can therefore evaluate any innovation in soybean disease control using their usual profit-maximization calculus. In this chapter, we analyze the farmer's decision to adopt an improved input for disease control with a decision tree model similar to the one we used in the previous chapter. In particular, we examine the case where a novel genetic trait offering broad resistance against oomycetes is developed and made available to farmers. This sort of disease resistance trait is not currently available in the market and, as such, the innovation is conjectural. It is not theoretical, however, as there is ongoing research pursuing this type of resistance (e.g. REEIS, 2017; Zhang *et al.*, 2017; Zhao *et al.*, 2017). Of course, even if the research proved immediately successful, the trait would be years away from commercialization. Our analysis is therefore forward looking but still instructive about how farm demand for innovation in soybean disease controls forms.

We begin by developing a standard decision tree model to examine the farmers' choice of whether to plant the new oomycete-resistant seed. The expected pay-off from the new seed is compared with those from other available control practices and the profit-maximizing choices determine the demand for the innovation. We then use sensitivity analysis to examine how this demand changes when the price of the new seed fluctuates. Next, we examine the potential market demand for such an innovation through a national survey of US soybean farmers. Because the innovation is not commercial and we cannot observe actual demand, we measure the willingness to pay (WTP) of farmers as an appropriate proxy. As before, we also survey expert CCAs to produce objective benchmarks and assess whether farmers' WTP (FWTP) estimates are consistent with those of experts as well as with the standard model of profit-maximizing behavior, as we would expect. Finally, we derive a projected market demand curve for the new oomycete-resistant seed and illustrate how adoption can be influenced by its market price.

The Economic Potential of Broad Genetic Resistance

The decision tree model we use here once again assumes that farmers make decisions for the control of seedling disease and mid-season root rots (specifically PRR) based on their expectations of disease incidence and severity, treatment method efficacy, loss abatement, and the cost of treatment inputs. Such decisions must be made early in the growing season, at the time of seed purchase, before the true nature of these conditions is known. Farmers can choose seeds with fungicide treatments, seeds with *Rps* genes, and seeds with the novel broad resistance trait to control early and mid-season oomycete diseases.

Economic value of broad resistance

The new trait is assumed to decrease the severity of both seedling disease and mid-season PRR by 90%, while costing $10.50/acre[29]. The specifics on the relative costs and efficacy of all other control practices as well as on the empirical distributions of disease incidence and severity, soybean prices, and other variables remain the same as those detailed in the previous chapter. We once again use Monte Carlo simulation with 1000 draws per model run. For the analysis, we considered three scenarios. In the first, the farmer only has to manage seedling disease. In the second, the farmer has to manage only mid-season PRR.[30] Finally, in the third, both seedling disease and PRR are to be managed.

The results of the baseline analysis are shown in Table 8.1. At the level of assumed efficacy, the broad resistance trait became the dominant disease control measure for all three scenarios examined. In particular, the new trait had a modest advantage in the seedling disease scenario, offering an additional value of $9.43/acre over the option of taking no action to control the disease,

Table 8.1. Optimal choices when broad resistance is available (Data from the USDA, industry, EMAC 2016 farmer survey, and author models.).

Disease control practice	Pay-off of disease control ($/acre)			Share of optimal solutions (%)		
	Seedling disease only	PRR only	Seedling disease + PRR	Seedling disease only	PRR only	Seedling disease + PRR
No action	–	–	–	10	4	0
Seed treatment only	6.23	–7.10	6.36	3	0	0
Rps genes only	1.62	6.69	12.18	0	30	0
Seed treatment + *Rps* genes	4.74	–0.41	15.63	0	0	0
Broad resistance trait	9.43	8.01	28.15	87	66	100

compared with a $6.23 pay-off for seed treatment alone. It was, however, the optimal choice for 87% of the soybean acres. The new resistance trait similarly had a modest advantage when PRR was to be managed, offering an additional value of $8.01/acre over the option of taking no action and compared with $6.69 returned by seeds with *Rps* genes alone. In this instance, use of the broad resistance trait was the optimal choice for 66% of acres. The full value of the novel broad resistance trait is realized when farmers must manage mid-season PRR in addition to seedling disease. Here, the value of the broad resistance trait, at $28.15/acre, was nearly twice that of the next best option, the combination of seed treatment and *Rps* genes, which provided a return of $15.63/acre over the option of no action. With this great value advantage, the broad resistance measure was the optimal choice for 100% of planted acres.

Price sensitivity analysis

The price at which the broad resistance trait would be offered in the market is an important factor that affects its pay-off and the farmer's profit-maximization calculus. For this sensitivity analysis, we added two price points, $7/acre and $14/acre, to the average price of $10.50/acre for further consideration. The pay-offs of the alternative disease control practices did not diverge from those in Table 8.1, as their costs and efficacies remained the same. In addition, the total revenue gain from the broad resistance trait remained unchanged, as its efficacy was held constant at 90%. However, the pay-off from the new trait did change and so did its adoption level. Figure 8.1 shows the inverse relationship between the per-acre pay-off of the new trait and its price per bag of seed.[31] As

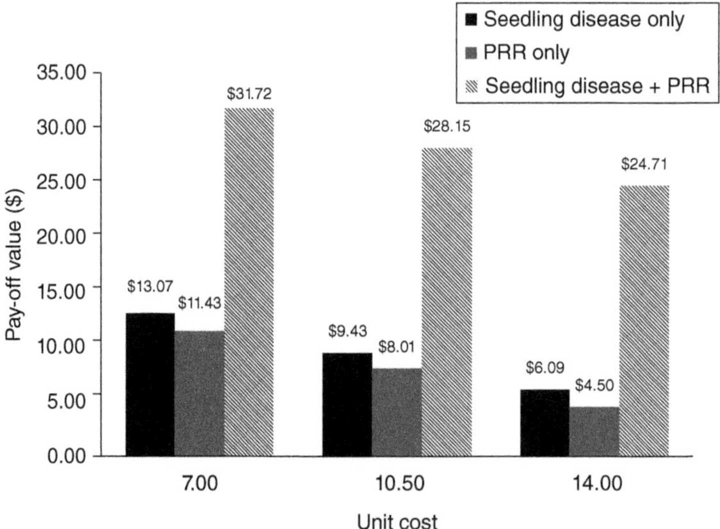

Fig. 8.1. Economic value of the broad resistance trait under different unit costs. (Data from EMAC farmer survey in 2016 and author models.)

the unit cost of the broad resistance trait increased, the pay-off derived in the control of seedling disease alone fell from $13.07/acre to only $6.09. Likewise, when the unit cost of the broad resistance trait increased from $7 to $14/acre, its pay-off in the control of PRR declined from $11.43 to $4.50. When both seedling disease and PRR are controlled, the pay-off of the novel trait fell from $31.72/acre to $24.71. The change in the economic value reflects the varying cost incurred by the farmer in order to obtain the same yield gains.

The changing price of the broad resistance trait also changed the farmer's acreage allocation among different disease controls. For instance, when only seedling disease is managed, at the highest unit cost of $14/acre, the broad resistance trait was the optimum choice in only 39% of acres. The less expensive option of seed treatment was preferred in 44% of acres, even though its efficacy is lower, while taking no action at all was optimal for 17% of acres (Fig. 8.2). Similarly, when only PRR is managed, at $14/acre the broad resistance trait was preferred in only 25% of acres while the less expensive option of seeds with *Rps* genes was optimal in 70% of acres. Finally, in the situation where a farmer managed mid-season PRR as well as seedling disease, the broad resistance trait maintained its superiority even at the highest price, due to its much higher efficacy. Even at $14/acre, the novel resistance trait was the optimal choice for 87% of acres while using seeds with *Rps* genes was optimal for 11% of acres.

Our simple sensitivity analysis above clarifies an important point that is often missed in technology adoption studies. Namely, that it is relative, not absolute, pay-offs that drive the adoption of improved practices and inputs. As the price of the novel broad resistance trait is increased, pay-offs decrease but remain positive. As the pay-offs fall below those of other control practices, adoption of the new trait wanes.

From the simple, stylized economic analysis above, it is clear that if a broad resistance trait against oomycetes were developed, it could see wide

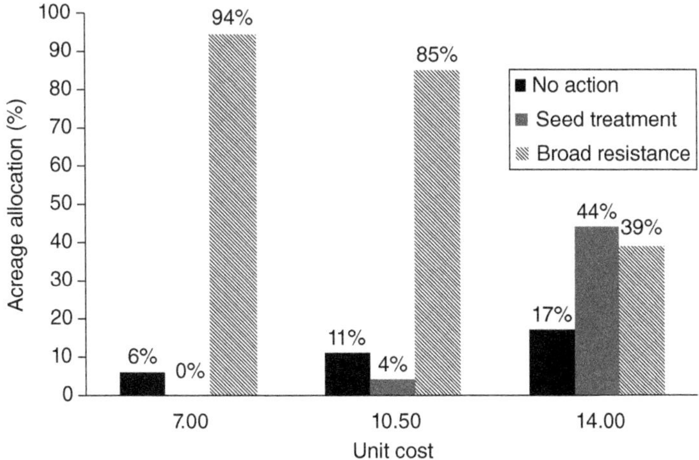

Fig. 8.2. Share of acres using the broad resistance trait under different unit costs compared with other treatments for seedling disease. (Data from EMAC farmer survey in 2016 and author models.)

adoption, especially by farmers who expect infestations both in the early point of the growing season and mid-season. Expected pay-offs from its adoption are higher than those from other commonly used control practices, making the new trait the profit-maximizing choice for the farmer's disease control consideration. Because of the inherent heterogeneity of farmers and farms, however, demand for the new trait would tend to vary across them. Of course, we cannot directly validate this result; we cannot observe actual farm demand as the new trait is not commercialized. Instead, we can evaluate FWTP as an appropriate proxy for potential demand and farmer interest.

Farmers' Willingness to Pay

WTP measures were initially developed to examine consumer product choices; Hahnemann (1991) provides a comprehensive discussion of the theory that underpins consumers' WTP. However, the concept applies equally to producers and firms (Lusk and Hudson, 2004). When a farmer considers the adoption of an improved input, the maximum amount of money the farmer would be willing to pay for the improved input is the difference in the farm's profits before and after adopting it. This result is derived whether the farmer's utility or profits are maximized (Zapata and Carpio, 2014). When FWTP is positive, it suggests some level of potential demand and willingness to adopt the improved input.

Measuring WTP

There are a few ways of collecting and analyzing WTP data; most involve surveys. Typically, respondents are presented with some representation of a consumer good or producer input and report their WTP some finite price for it. Contingent valuation is probably the most commonly stated preference research approach (e.g. Hahnemann, 1991; Lusk and Hudson, 2004). Data is collected via surveys administered in person, by mail, or online, so the method is relatively inexpensive and can handle hundreds or thousands of respondents. Contingent valuation data can be collected through different types of survey approaches:

- *Open-ended questions* ask respondents to directly state a specific, monetary value they are willing to pay for a described product. This type of approach elicits the most detailed information from each respondent. It also simplifies the analysis, as average WTP can be calculated arithmetically from all responses. However, open-ended questions have been shown to have an increased risk of bias from extreme responses, either high or low.
- *Single-bounded dichotomous choice questions* present the respondent with both a described good and a proposed price and ask only if the respondent would buy the good or input at that price or not. The proposed price is varied systematically to get a distribution of responses. This approach places a much simpler cognitive task on the respondents and

eliminates the possibility of outliers. This comes at the expense of detail, however. Average WTP is estimated statistically from the distribution of responses.

- *Double-bounded dichotomous choice questions* begin as a single-bounded dichotomous choice, but then add a follow-up question. If respondents indicate that they would (would not) buy at the initial price, they are given a higher (lower) price for the same good and again asked if they would buy. These items elicit more detailed information from each respondent and the possible magnitude of individual WTP is restricted to a smaller range by the follow-up question. Average WTP must still be estimated by statistical methods.

Each of these approaches has its own advantages and disadvantages. In addition to average WTP, each method can also provide a projected demand curve. This can be done directly with data from open-ended choice questions, but again only through statistical methods with dichotomous choice data.

FWTP studies

The concept of WTP has been used empirically to examine farmer demand for novel and improved inputs. For instance, Hudson and Hite (2003) used a mailed open-ended contingent valuation survey to investigate FWTP for a site-specific field management technology package. Marra *et al.* (2010) used a single-bounded dichotomous choice design in a mailed survey to elicit the WTP of farmers in the southern USA for a yield monitor designed for cotton production. Hubbell *et al.* (2000) combined revealed farmer preference data from actual sales of insect-resistant cotton seed in its first year of commercialization with stated preference data gathered by a single-bounded contingent valuation survey of non-adopters. The two levels of data amounted to a double-bounded contingent valuation design and were used to estimate WTP and demand curves for prices below the observed market price. Qaim and de Janvry (2003) used a similar approach to investigate FWTP and derive demand curves for insect-resistant cotton seed in Argentina. Finally, Matuschke *et al.* (2007) examined the adoption of hybrid wheat in India using the same type of double-bounded contingent valuation design as the previous two studies.

Soybean Farmers' Willingness to Pay for Broad Disease Resistance

For our study, we used our national surveys, the first in 2013 and the second in 2016, in order to evaluate the WTP of US soybean farmers for a broad resistance trait against oomycetes. We also surveyed expert CCAs about their perceptions of what soybean farmers might be willing to pay for such a novel trait in 2012 and again in 2015. Farmers often consult with CCAs for advice in seed selection decisions. This gives the experts knowledge and insight about farmer preferences and potential WTP.

We used open-ended questions for our contingent valuation surveys of the CCAs and for one of the farmer surveys. For the other farmer survey, we used a double-bounded dichotomous choice instrument.[32] Our surveys elicited FWTP for the novel broad resistance trait in seedling disease control and in PRR control, separately. As such, in our surveys we asked the soybean farmers: "If a new gene that provided total resistance to seedling disease (mid-season root rot) was found, which eliminated the chance of damage from this disease in your fields and this gene/trait was bred into the elite varieties you plant, what maximum premium per bag of seed you would be willing to pay for that trait?" Similarly, experts were asked: "If a seed variety that is broadly and effectively resistant to seedling disease (mid-season root rot) were developed, what maximum premium per bag of seed do you think farmers in your area would be willing to pay for that seed? Please assume that the seed variety would carry elite yield and other traits important in your local area."

Average FWTP, as perceived by expert CCAs, is reported in Table 8.2. FWTP for the new resistance trait in seedling disease control in 2012 was, on average, $7.27/acre, while in 2015 it was $5.84. The expert perceptions of FWTP for the new trait for PRR control were similar, although the values were somewhat lower. In 2012, the experts believed that farmers were willing to pay $5.66/acre if PRR could be eliminated and in 2015 that figure was $4.72. Overall, the CCAs expected FWTP to be lower in 2015 than in 2012, a result that is consistent with observed market conditions as soybean prices declined over this 3-year period.

In comparison with the expert perceptions of FWTP, in 2013 farmers stated they were, on average, willing to pay $8.68/acre for seedling disease resistance and $8.53/acre for PRR resistance. In 2016, US soybean farmers reported average WTP of $7.22/acre and $5.64/acre for seedling disease and PRR resistance, respectively, probably in response to the lower soybean prices. Thus, overall, farmer WTP values were similar to those expected by the experts.

When these valuations are broken down by state, the WTP for seedling disease and PRR resistance suggested by farmers and experts (Figs 8.3 and 8.4) were also comparable. With a few exceptions, the states with the highest incidence and severity of disease had higher valuations. It is worth noting that the farmer surveys show greater variability in the WTP values than the CCA surveys. The more limited sample size of the farmer surveys, especially in

Table 8.2. Average farmer willingness to pay for broad resistance trait (Data from EMAC 2016 farmer survey and author models.).

Year	Seedling disease ($/acre)	PRR ($/acre)
CCA surveys		
2012	7.27	5.66
2015	5.84	4.72
Farmer surveys		
2013	8.68	8.53
2016	7.22	5.64

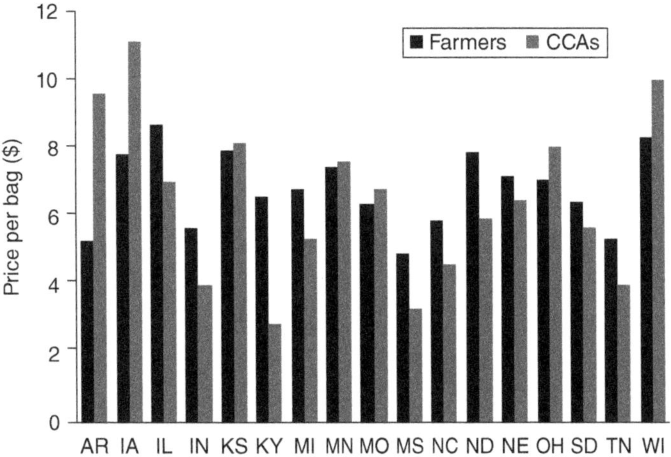

Fig. 8.3. WTP for the new seedling disease resistance trait by state. (Data from EMAC farmer surveys in 2013 and 2016, EMAC CCA surveys in 2012 and 2015, and author models.)

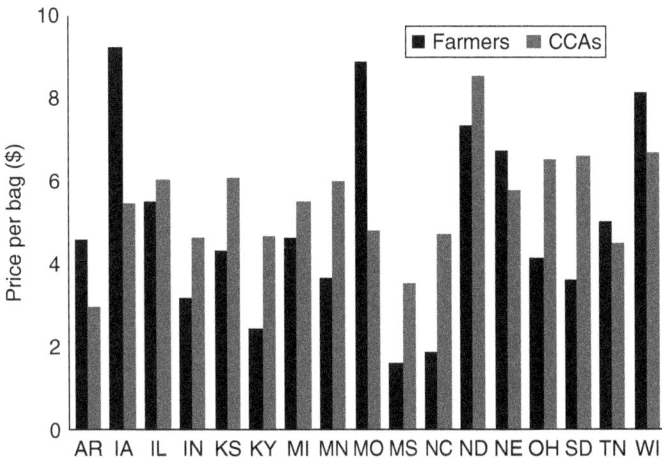

Fig. 8.4. WTP for the new PRR resistance trait by state. (Data from EMAC farmer surveys in 2013 and 2016, EMAC CCA surveys in 2012 and 2015, and author models.)

states with limited soybean production (e.g. Kentucky, Mississippi, Tennessee), allowed individual responses to have a stronger effect on the reported averages.

The distribution of WTP across respondents is also telling. The distributions of the CCA-perceived FWTP for seedling disease control in 2012 and 2015 are illustrated in Fig. 8.5. Here, only a small share of the experts thought that farmers would pay nothing. However, there was a relatively wide range of FWTP values that the experts perceived, with the bulk of responses being between $1 and $12. The largest share of experts, over 30%, believed that FWTP was between $4/acre and $6/acre.

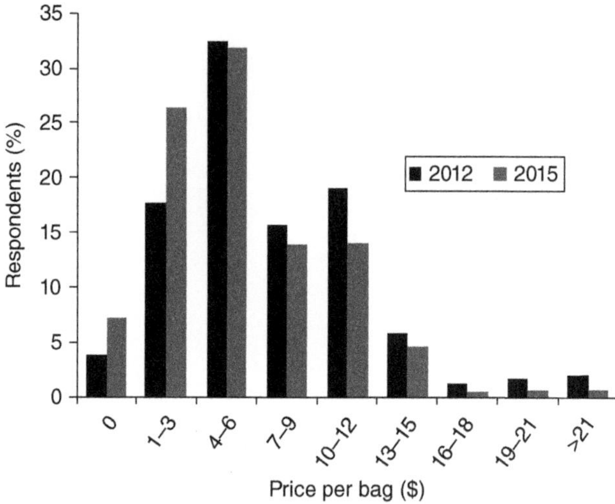

Fig. 8.5. Distribution of CCA estimates of FWTP for the seedling disease resistance trait. (Data from EMAC farmer surveys in 2013 and 2016, EMAC CCA surveys in 2012 and 2015, and author models.)

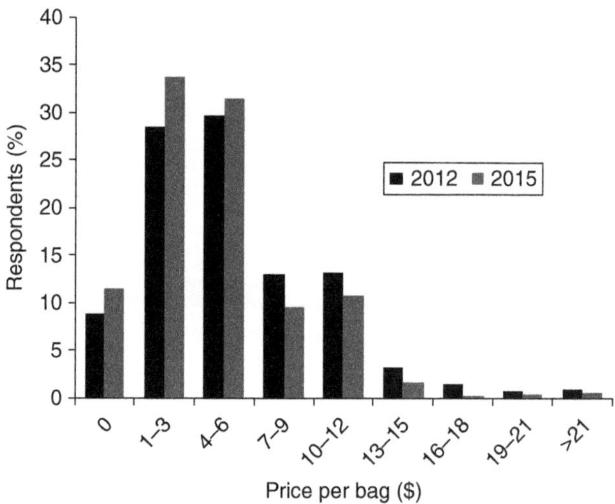

Fig. 8.6. Distribution of CCA estimates of FWTP for the mid-season root rot resistance trait. (Data from EMAC farmer surveys in 2013 and 2016, EMAC CCA surveys in 2012 and 2015, and author models.)

The distribution of experts' perceived FWTP for PRR resistance is shown in Fig. 8.6. Roughly 10% of experts believed farmers would pay nothing, but the majority of experts believed farmers would pay between $1/acre and $6/acre. Overall, the distributions of FWTP as perceived by CCAs were consistent over both the 2012 and 2015 surveys.

The distributions of FWTP values for the broad resistance against seedling disease and PRR are equally informative. Figure 8.7 shows that the majority of

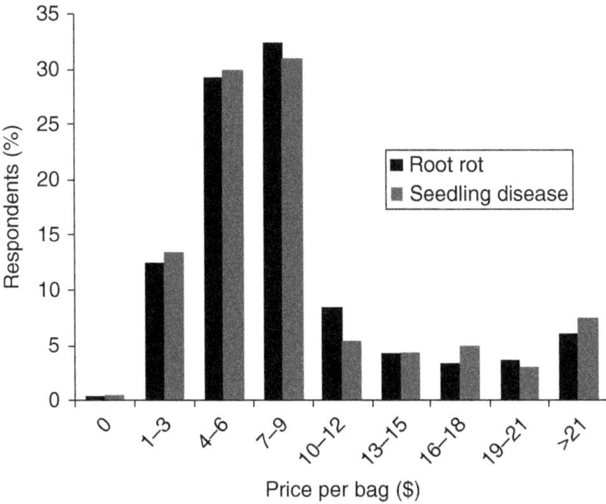

Fig. 8.7. FWTP for the broad resistance trait. (Data from EMAC farmer survey in 2013 and author models.)

the soybean farmers were willing to pay between $4/acre and $9/acre. Roughly 25% of farmers were willing to pay more and 12% would pay less. Overall, we found that FWTP estimates, although marginally higher, were consistent with those of the expert CCAs in magnitude and distribution.

Factors Affecting Farmers' Willingness to Pay

With these detailed data in hand, it is also possible to examine the factors that shape FWTP for the novel broad resistance trait against oomycetes. Earlier in this chapter, and within the context of the decision tree model, we discussed how farm demand for innovation in disease control is shaped by the farmer's standard profit-maximizing decision process and illustrated how the farmer's expectations of disease incidence and severity as well as the relative efficacy of control practices determine such demand. We therefore expect these same factors to condition FWTP for the broad resistance trait. Because farmers and farms are heterogeneous, their optimal disease control decisions may differ and hence farmer and farm characteristics may also influence FWTP for the novel trait.

In this context, we estimated two statistical models where the dependent variables were FWTP for seedling disease and PRR control, respectively. Twelve independent variables were used in each of these two models, representing farmers' expectations and farmer/farm characteristics (Table 8.3).

Several indicators of farmer characteristics were included in the statistical analysis. Farmers who identify themselves as early adopters of agricultural innovations (ADOPT) may consistently value new technologies more highly than other farmers. Farmers with more experience in soybean production (YEARS) might also be in a better position to evaluate the economics of a new

Table 8.3. FWTP models: description of explanatory variables and survey questions.

Variable name	Question
ADOPT	Which of the following best describes your adoption of new products or farming practices? Would you say you are generally: among the [first/middle/last] one-third of growers in your area to try and adopt a new product or farming practice?
ACRES	How many acres of soybeans did you plant this year?
YEARS	Including 2012, for how many years have you been growing soybeans?
YIELD	Estimated yield impact of disease (bushels/acre)
EARLY	Based on your experience, what is the economic advantage, if any, of planting soybeans earlier in the season as opposed to later in the season, i.e. how many additional bushels/acre, if any, do you think you get from earlier-planted soybeans as opposed to later-planted soybeans?
NOTILL	Based on your experience, what is the economic advantage, if any, of a no-till soybean system, i.e. how many additional bushels/acre, if any, do you think you get from a no-till soybean system compared with conventional tillage? How much in terms of $/acre, if any, do you think you save on input costs with a no-till soybean system compared with conventional tillage?
TREAT	In the last 5 years, on average, what percentage, if any, of the soybean seed you bought was treated with fungicides?
RESIST	In the last 5 years, on average, what percentage, if any, of the soybean seed you bought was selected because the varieties were resistant to [seedling disease/*Phytophthora*]?
ROTATE	Based on what you know from your fields but also from what you have seen or heard from other farmers and experts in your area, how effective is rotating to another crop in order to avoid [seedling disease/*Phytophthora* root rot later in the season]?
TILE	Based on what you know from your fields but also from what you have seen or heard from other farmers and experts in your area, how effective is tiling the field in order to avoid [seedling disease/*Phytophthora* root rot later in the season]?
INSURANCE	To what extent, if at all, would you agree that crop insurance offers "peace of mind" when it comes to soybean stand establishment problems or losses from mid- and late-season diseases?
TEST	If a simple, quick, and low-cost test to identify seedling disease pathogens in the field were available, how likely would you be to use it in your soybean fields?

technology, such as the broad resistance trait. Finally, farmers who might feel less knowledgeable and certain about the incidence of disease on their farm might be more inclined to test (TEST) in order to identify disease pathogens and could exhibit higher WTP for broad resistance.

Farm size can sometimes affect farm demand for innovation and technology adoption, but the direction of the overall impact can be ambiguous in the case of disease control. Larger farm operations may have greater financial resources to pay for innovation but smaller farm operations might be more intensively cropped; as such, both could have a positive impact on FWTP for the broad resistance trait. We have therefore included the number of soybean acres cropped (ACRES) as an indicator of farm size, but we make no assumption on the direction of the overall impact ACRES might have on FWTP.

Farmers' expectations of the incidence and severity of seedling disease and mid-season root rots would be expected to increase FWTP for the broad resistance trait. As the value of any disease control measure is based on the loss it prevents, we asked farmers to estimate the yield loss due to seedling disease and PRR that they typically experienced on their farm (YIELD).

FWTP for the broad resistance trait should be, in part, determined by its expected relative efficacy. The farmers participating in our survey, however, did not have any such prior expectations. In fact, we defined the potential efficacy of the novel trait in the survey as part of our description of the technology. For this reason, we did not include in the model farmers' expectations of relative efficacy for the new technology and for other control practices that directly compete with it (e.g. partial resistance and seed treatments). Instead, we used two other related proxies: the share of the seeds farmers buy that were treated with fungicides (TREAT), and the share of seeds with partial resistance or *Rps* genes (RESIST). Substantial farm use of either of these practices would indicate an elevated need for control and a positive effect on FWTP for the broad resistance trait. High values of RESIST may also indicate farmers' familiarity with genetic resistance in the seed and hence have a further positive effect on FWTP for the novel trait.

Several agronomic practices can affect disease but can also influence the overall farm profitability through yield and cost-efficiencies. Earlier planting dates (EARLY) and conservation tillage (NOTILL) both offer yield and cost advantages but can also increase the chance of oomycete incidence and severity. The relative magnitude of such benefits compared with any yield losses from disease, as perceived by individual farmers, would be expected to influence their WTP. In order to measure these two variables, we added the perceived yield and cost benefits of these two practices in monetary terms. We therefore expect a positive impact of NOTILL on FWTP for seedling disease or PRR control and a positive impact of EARLY on FWTP for seedling disease control alone, as early planting does not affect mid-season root rots.

Farmers were asked to evaluate the potential effectiveness of crop rotation (ROTATE) and tiling fields to improve drainage (TILE) in avoiding oomycete diseases. A high expected efficacy for such practices could have a negative impact on FWTP for the broad disease resistance trait. Other farm management practices could also influence FWTP for the broad resistance trait. Crop insurance (INSURANCE) provides a backstop to limit losses from varying degrees of crop failure, and thus may be somewhat of a substitute for disease control measures, including the broad resistance trait.

Based on these considerations, we specified two regression models, the first for seedling disease control and the second for mid-season root rot control, using the following general specification:

$$WTP = \beta_0 + \beta_1 ADOPT + \beta_2 ACRES + \beta_3 YEARS + \beta_4 INSURANCE$$
$$+ \beta_5 YIELD + \beta_6 EARLY + \beta_7 NOTILL + \beta_8 TREAT + \beta_9 ROTATE$$
$$+ \beta_{10} TILE + \beta_{11} RESIST + \beta_{12} TEST + \varepsilon \tag{8.1}$$

As the dependent variables are discrete, ordinary least squares (OLS) estimation is not appropriate because some basic assumptions of OLS are violated (Pindyck and Rubinfeld, 1981). Accordingly, the coefficients in the above equation were estimated by means of an ordered multinomial logit model and the estimated parameters are reported in Table 8.4.

In both models, indicators of the expected incidence and severity of disease had the most significant impact. Farmers' expectations of the yield impact of disease (YIELD) had a large and statistically significant impact on FWTP for both seedling disease and PRR resistance. In both cases, a greater perceived yield loss to disease led farmers to place a higher value on the new broad resistance technology. The share of existing genetic resistance in the seed that farmers buy also had a similarly large and statistically significant impact on FWTP for the new trait.

The perceived economic benefit of no-till also had a statistically significant and positive impact on farmers' valuation of oomycete control. This suggests that farmers who practice no-till are aware of its potential impact on disease, even though they see a net yield benefit and added cost-efficiencies. Hence, use of no-till practices on the farm has a positive impact on FWTP for broad resistance that can mitigate any negative impact on disease. Earlier planting dates, which can also lead to a higher incidence of disease, had a positive and statistically significant impact on FWTP for seedling disease control but, appropriately,

Table 8.4. Factors affecting FWTP for the broad resistance trait. (Data from EMAC 2013 farmer survey and author models.)

Variable	Mid-season root rot	Seedling disease
Producer is early adopter of new tech (ADOPT)	2.0133*	2.4927*
Soy acres planted (ACRES)	−0.0017	−0.0019*
Years producing soybeans (YEARS)	−0.0687	0.0399
Estimated yield impact of disease (YIELD) (bushels/acre)	0.6028***	0.1709*
Estimated benefit of no-till (NOTILL) (bushels/acre)	3.8868**	2.981*
Estimated benefit of early planting (EARLY) (bushels/acre)	−0.0257	0.2609*
Share of seeds bought with partial resistance and *Rps* traits (RESIST)	0.0203**	0.0385**
Share of seeds bought with fungicide treatments (TREAT)	−0.0838	1.478
Perceived effectiveness of crop rotation (ROTATE)	1.9309	1.237
Perceived effectiveness of improving drainage (TILE)	2.6415	3.5276
Insurance offers "peace of mind" (INSURANCE)	3.852	0.4454
Farmer would use test to identify causal disease (TEST)	1.5429*	2.7793***

***, Statistically significant at the 1% level; **, statistically significant at the 5% level; *, statistically significant at the 10% level.

not for PRR control. Early planting primarily affects the incidence of seedling disease and not that of mid-season root rots.

Some of the individual farmer characteristics also had an important impact on the valuation of the technology. Farmers who self-identified as "early adopters of technology" were more likely to place a higher value on the new genetic resistance trait for both seedling disease and PRR control. Similarly, farmers who were more likely to test for disease on their farm also placed a higher value on the new broad resistance trait. The remaining independent variables we used in the regression models did not have statistically significant impacts on FWTP for the novel broad resistance trait.

In sum, from our statistical analysis, we find that farmers' expectations about disease incidence and severity on their farm tend to have the most significant impact on their WTP for the improved control input. Farmers' expectations of the benefits and costs of other control practices (e.g. early planting and no-till) as well as their personal characteristics can also affect their WTP for the novel broad resistance trait. Thus, overall, our results suggest that FWTP for the new resistance trait is consistent with the standard profit-maximization decision process we have discussed throughout this book.

Potential Market Demand for the Broad Resistance Trait

Using FWTP for the broad resistance trait against oomycetes, we can also develop demand curves that project the potential level of adoption of the novel trait against a price schedule. For instance, in the constructed demand curve shown in Fig. 8.8, each point on the line represents the proportion of respondents who are reportedly willing to pay a given price for broad resistance against seedling disease.

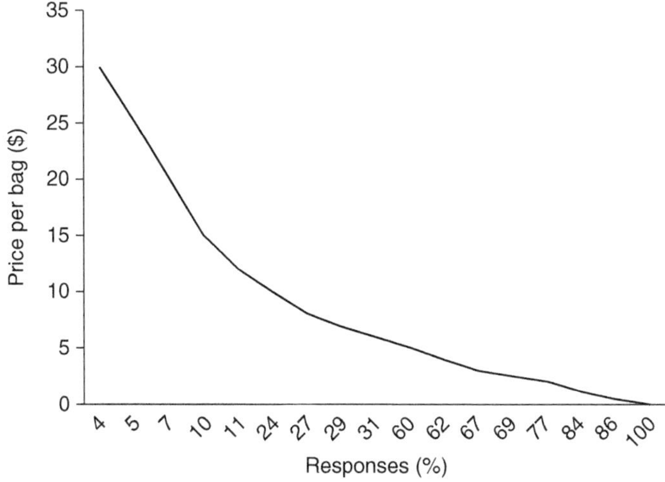

Fig. 8.8. Projected demand curve for broad resistance against seedling disease. (Data from EMAC farmer survey in 2013 and author models.)

As we discussed above, the WTP of soybean farmers is shaped by their individual expectations of disease incidence and severity and the cost and effectiveness of alternative control measures and the novel trait, as well as other individual factors. These expectations determine the economic value (the expected pay-off) of the novel trait for each farm and, hence, FWPT. Farmers who expect a high incidence and severity of seedling disease on their farm, for instance, would expect high pay-offs from the use of the broad resistance trait, and they would tend to express a high WTP for it. The price of the broad resistance trait also shapes its expected pay-off for any given farm. Thus. at different price levels, the proportion of soybean farmers who find the broad resistance trait to be the profit-maximizing solution and who are willing to pay this price define the potential market demand and level of adoption of the novel trait.

Using the projected demand in Fig. 8.8, we can see that if, for example, the price of the novel resistance trait was set at $7 per bag of seed, 29% of the US soybean farmers would be willing to pay for it and use it to control seedling disease. This is approximately the same proportion of US soybean farmers reporting seedling disease on their farm, and hence the constructed potential demand curve suggests that, at $7/acre, farmers experiencing seedling disease would adopt the broad resistance trait for effective control. At $14/acre, however, only 10% of US soybean farmers would be willing to pay for the novel trait, indicating a much lower potential demand and adoption.[33]

The projected demand curve for the broad disease trait in Fig. 8.8 also illustrates another important consideration in the adoption of improved practices on the farm. At a $7/acre cost for the new trait, 71% of US soybean farmers are not adopters, but the WTP for the majority is not zero. This latent demand for the novel trait can be turned into actual market demand if the novel trait is priced lower. As Fig. 8.8 illustrates, at a $3/acre cost, for example, 67% of US soybean farmers would be willing to pay and adopt the broad resistance trait. At this lower cost, the new trait offers insurance and economic value even for farmers who may not experience strong pressure from the disease. How the suppliers choose to price innovations in disease control inputs will therefore tend to influence their level of overall adoption.

Summary

Presented with the possibility of a novel broad genetic resistance trait against oomycetes, soybean farmers demonstrate WTP for and willingness to adopt the trait comparable to those expected by experts and consistent with the standard profit-maximizing calculus. Indeed, the damage abatement and standard profit-maximizing decision-making process underpin farmers' demand for any innovation in disease control. Farmers form expectations of disease incidence and severity, treatment method efficacy, loss abatement, and the costs of treatment for all alternative control practices and inputs, including the novel one. They then compare the expected pay-off from the innovation with those from the alternatives. When the innovation is the profit-maximizing option, they adopt it. The relative pay-off of the innovation is its basic economic value, which limits the soybean FWTP. Because farmers and farms are heterogeneous, they arrive at different profit-maximizing choices and hence have differing WTP for any innovation in disease control. The distribution of the WTP of all farmers defines the market demand for the innovation. How suppliers choose to price the innovation can strongly influence its market demand and level of adoption, a topic we address further in the next chapter.

Notes

[29] We interviewed executives in the seed industry about the potential price premium that soybean farmers with seedling disease and mid-season PRR incidence in their fields might be willing to pay for a broad resistance trait: $10.50 was the mid-point of such estimates.

[30] The first two scenarios therefore represent the sort of disease control decisions made by farmer groups that experience seedling disease or mid-season root rots on their farms, roughly a third of all soybean farmers in the USA (see Chapter 5, this volume).

[31] The terms 'price per bag' and 'cost per acre' for the broad resistance trait are used interchangeably here as the average seeding rate for soybeans in the USA is one bag of seed per acre.

[32] The first farmer survey was carried out by telephone and the interviewer asked the participants if they were willing to pay a random price chosen from a range of likely prices. If the farmer replied negatively, the administrator repeated the question using a price that was $3 lower. Conversely, if the farmer replied positively to the initial question, then the administrator repeated the question with a $3 higher price.

[33] Note that these figures are also consistent with the results we derived from the decision tree analysis presented earlier in this chapter. As illustrated in Fig. 8.2, at $7/acre, 94% of the acres of the farmer segment experiencing seedling disease (about a third of all US soybean farmers) would use the broad resistance trait. At $14/acre, however, the trait would be used on only 39% of the acres of this farmer segment. If these averages were generalized, at $7/acre the overall market adoption of the broad resistance trait would be almost 32% of US soybean acres (94% of 34% of total soybean acres) and at $14/acre, overall market adoption would be approximately 13% of US soybean acres (39% of 34% of total soybean acres).

Supply of Inputs for Disease Control

When making decisions about disease control in any given growing season, farmers' choices for prevention and treatment are often limited by the decisions of scientists, engineers, and managers made many years in advance. Inputs such as seed treatments and genetic resistance are supplied by commercial crop protection and genetics firms that invest heavily in R&D. The collective economic decisions of such input suppliers shape the supply of inputs for soybean disease control and are the focus of this chapter.

We begin by examining the physical, technical, regulatory, and market conditions that have affected the discovery and development of various chemical controls and genetic resistance traits used to combat soybean disease. Next, we look at the existing and future pipelines of such inputs and draw conclusions about the overall supply of disease control inputs available to soybean farmers now and in the future. As in previous chapters, we focus on chemical controls and genetic technologies developed to combat seedling disease and mid-season root rots when gaining insight from detail is necessary.

Chemical Controls: Research and Development

In the last 70 years, there has been an active and ongoing search for compounds that will control pests without damaging crops or the environment. During this period, development of pest resistance to some deployed compounds created the need for replacements. Regulatory requirements for crop protection products also became progressively more stringent, with added demands on environmental, non-target, and toxicological product profiles. As a result, crop protection firms have spent more time and effort searching for new active ingredients with improved efficacy and selectivity that could also meet more stringent regulatory requirements. In practice, this has meant screening more molecules to find new marketable active ingredients. Ongoing development of new research methods and tools has enabled the crop protection industry to evaluate an ever-increasing number of molecules.

Technology

Crop protection product discovery is based on the synthesis and mass screening of large numbers of chemical compounds. The traditional approaches for discovery and screening of pesticides have included empirical synthesis with conventional screening, analog synthesis, natural product models, biochemical design and biorational synthesis. Such traditional methods of chemical synthesis were relatively low throughput. For example, in the 1960s and 1970s, an experimental chemist using such methods might have been able to produce 100 target compounds for screening per year (Ridley *et al.*, 1998).

In more recent years, advances in combinatorial chemistry and high-throughput screening have enabled drastic increases in the number of compounds being evaluated (Sparks, 2013; Sparks and Lorsbach, 2016). The main objective of combinatorial chemistry is to rapidly synthesize many compounds using a process that is supported by computation and automation. Through the use of robotics, data processing and control software, liquid handling devices, sensitive detectors, and other foundational tools, high-throughput screening allows researchers to quickly conduct millions of chemical, genetic, or pharmacological tests. Researchers can then more rapidly identify active compounds that modulate particular biomolecular pathways. The results of these experiments provide starting points for chemical design and for understanding the interaction or role of a particular biochemical process. Nevertheless, only a very small number of all of the compounds initially identified through high-throughput screening are ultimately developed.

More targeted approaches for generating and evaluating leads have also been developed in order to improve upon random screening. Until recently, the dominant paradigm for the discovery of agrochemicals involved a "chemistry-first" approach (Tietjen and Schreier, 2012). This was largely a function of the need to test the chemicals *in vivo* (Lamberth *et al.*, 2013). Advances in genomics and bioinformatics have allowed the identification of potential pesticide targets at the molecular level. With the fields of molecular and structural biology growing rapidly, computer-aided molecular design has become a preferred approach. Here, it is possible to rationalize the experimental data *in silico* in order to guide the design of new pesticides.

Computer-aided pesticide design involves many complementary analytical methods (e.g. Benfenati, 2007; Speck-Planche *et al.*, 2011; Zhang, 2011). For instance, virtual screening is a structural analytical approach used by computational chemists and biologists for the initial screening of virtual organic molecules in order to arrive at a manageable number of compounds to synthesize and evaluate, after it has been confirmed that they can bind to targets of interest and are thus likely to yield feasible pesticide candidates. Other analytical methods include homology modeling, molecular docking, and molecular dynamics simulation. Quantitative structure–activity relationship (QSAR) models relate the biological activity of a molecule to some selected features of their physico-chemical structure by means of statistical tools. QSAR models are based on the assumption that the biological activity of a compound is related to its molecular properties. Computer-aided molecular design of pesticides is sometimes referred to as structure-based design and relies on knowledge of the three-dimensional structure of the biomolecular target (Walter, 2002; Saini and Kumar, 2014). This contrasts with more conventional approaches that focus on the design of a molecule that will bind to its target.

In silico or virtual approaches increase the likelihood of finding some relevant pesticidal activity with comparatively fewer compounds synthesized and screened (Drewes *et al.*, 2012). There is also evidence that the use of *in silico* testing and virtual filters may be having an impact on the number of products per screen (Sparks and Lorsbach, 2016). However, the downside to such *in silico* tools is that they may also filter out unusual or unexpected compounds. As a result, most crop protection firms that develop new active ingredients employ a range of approaches (Shelton and Lahm, 2015; Loso *et al.*, 2016). The integration of many innovative technologies and tools now allows a "from genome to pesticide" approach in crop protection R&D (Saini and Kumar, 2014). It is worth noting that many of the hardware and software tools of computer-aided pesticide design were originally developed in the pharmaceutical industry (Benfenati, 2007) and then transferred to agrochemistry (Scherkenbeck, 2009).

Pest resistance and regulatory requirements

The development of new research methods and tools has enabled the crop protection industry to evaluate an ever-growing number of molecules in order to meet its expanding demands. Dynamic pest populations and their ongoing development of resistance to existing pesticides has been a key factor driving the continual need for new active ingredients. Across the globe, an increasing number of pathogens, as well as insects and weeds, has been reported to demonstrate resistance to various chemistries (Fig. 9.1).

Reports of pest resistance to chemical controls date from the early 20th century (Retzinger and Mallory-Smith, 1997; Sparks and Nauen, 2015). However, the number of resistant species and the rate at which newly resistant species are appearing, particularly among pathogens and weeds, appear to be increasing over time. A number of factors contribute to the development of resistance, including the reproductive biology and ecology of the pest and the frequency and intensity of pesticide application (Whalon *et al.*, 2008). Resistance is most closely associated with the intensive and exclusive use of one pesticide or a small group of pesticides with the same or similar modes of action. Such a pattern of use places pest populations under focused selection pressure that greatly increases the probability of resistance (Powles, 2008). Resistance closes off known target sites, requiring novel modes of action. Furthermore, these novel modes of action must be scrutinized for their durability to resistance. As a result,

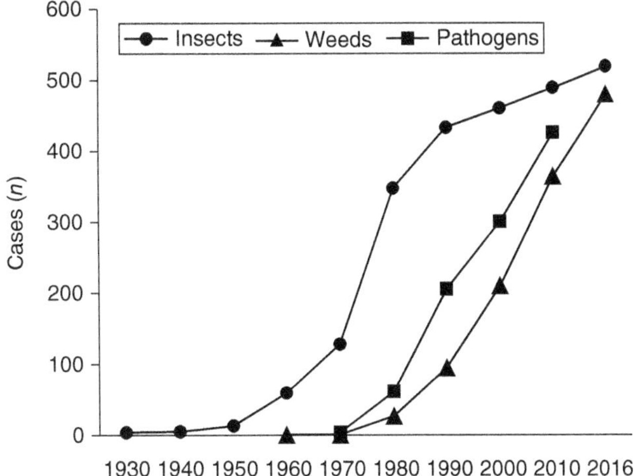

Fig. 9.1. Pest resistance build-up over time. (Data from www.frac.info/, www. pesticideresistance.org, www.weedscience.org and the authors.)

resistance considerations have added to the number of compounds that must be screened in order to find suitable candidates.

Stricter regulatory requirements have also increased the number of compounds that must be screened by the crop protection industry. The Food Quality Protection Act, passed in 1996, and other laws and regulations installed in the USA and other countries led to many older chemistries losing their registration and many more new chemistries being evaluated in order to identify those few that could meet the more stringent toxicological and environmental standards required (National Research Council, 2000; Sparks and Lorsbach, 2016).

Given these conditions, the crop protection industry has been screening an ever-increasing number of compounds in its R&D programs. In the 1950s, approximately 1800 compounds were screened per product commercialized, increasing to 10,000 compounds by 1972 (Menn, 1980; Metcalf, 1980). By 1977, the number of compounds screened per product commercialized had risen to 20,000 (Menn, 1980), increasing to 50,000 by 1990–1994 (Stetter, 1993). More recently, in 2009–2014, the number was estimated to be 160,000 compounds screened per product discovered (Phillips McDougall, 2016b). Despite the increased flow of screened molecules in recent years, however, success in discovery has steadily diminished as the number of new active ingredients coming to the market has declined. In all classes of crop protection chemicals, there has been a definite downward trend in product introductions over the last 15–20 years that seems to be accelerating. Introductions since 2011 have fallen even further off the pace of recent decades (Fig. 9.2).

As the number of new active ingredients coming to the market has declined, the amount of time as well as R&D and regulatory compliance spending required to commercialize a new active ingredient have increased in parallel. It is

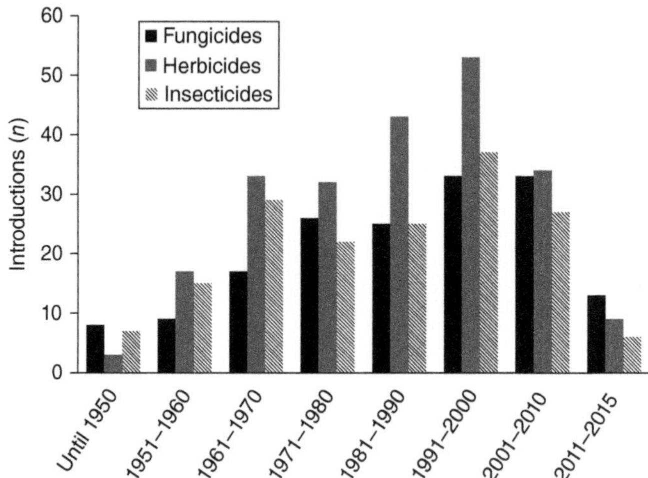

Fig. 9.2. New active ingredient introductions. (Data from Phillips McDougall, 2016b.)

generally recognized that moving a new pesticide through the development, regulatory approval, and registration process is expensive, costing as much as $286 million (Phillips McDougall, 2016b). Furthermore, the process has long lead times, requiring anywhere from 8 to 12 years for completion (National Research Council, 2000; Holm and Baron, 2002). Estimates suggest that the development, regulatory approval, and registration process is gradually becoming longer on average, increasing from 8.3 years in 1995 to 9.8 years in 2005–2008 and to 11.3 years in 2009–2014 (Phillips McDougall, 2016b).

Chemical Controls for Soybean Disease

Crop protection firms are willing to spend large sums on R&D because the rewards can be large as well. The global market for chemical crop protection products was valued at $51.2 billion in 2015, with fungicides accounting for 27% of such spending (Phillips McDougall, 2016a). Currently, there are 165 active ingredients from 13 chemical categories registered as agricultural fungicides (Table 9.1; Phillips McDougall, 2016a), but a much smaller number of them account for most of the use around the globe. Strobilurins, sterol biosynthesis inhibitors, and succinate dehydrogenase inhibitors together account for almost 70% of all fungicides sales (Table 9.2). Even within each class, the use of active ingredients is quite concentrated. For example, of 46 sterol biosynthesis inhibitor fungicides, the top six account for over 60% of the market. Of the 12 strobilurin active ingredients, the top three command nearly 80% of sales (Phillips McDougall, 2016a). As such, the unique modes of action of fungicides used in agriculture are few.

Soybeans account for 15% of the global fungicide market, in third place behind grains (roughly 30%) and fruits and vegetables (about 25%) (Phillips

Table 9.1. Fungicide classes and modes of action. (Data from Phillips McDougall, 2016a.)

Chemical class	Mode of action
Anilinopyrimidines	Inhibition of methionine biosynthesis Inhibition of secretion of hydrolytic enzymes
Benzimidazoles	Inhibition of β-tubulin synthesis
Dicarboxamides	Multisite contact Inhibition of triglyceride synthesis
Dithiocarbamates	Multisite contact
Inorganics	Multisite contact
Morpholines	$\Delta 8$ to $\Delta 7$ isomerase and $\Delta 14$ reductase inhibitors
Phenylamides	Interference with ribosomal RNA causing inhibition of protein synthesis
Phthalimides/phthalonitriles	Multisite contact
Pyrimidines	C14 demethylase inhibition in the sterol biosynthesis pathway
Strobilurins	Inhibition of mitochondrial synthesis at the cytochrome bc_1 complex
Succinate dehydrogenase inhibitors	Inhibition of succinic acid oxidation during metabolic respiration
Triazoles	C14 demethylase inhibition in the sterol biosynthesis pathway
Other azoles	C14 demethylase inhibition in the sterol biosynthesis pathway

Table 9.2. Fungicide sales by class. (Data from Phillips McDougall, 2016a.)

Chemical class	Sales ($) 2010	2015	2020
Anilinopyrimidine	215	242	270
Benzimidazole	441	441	455
Dicarboxamide	170	163	160
Phenylamide	350	384	410
Strobilurin	2762	3579	3890
Succinate dehydrogenase inhibitors	428	1576	2200
Multisite			
Dithiocarbamate	800	862	1,015
Inorganic	665	850	955
Phthalimide	450	490	540
Other	320	305	335
Sterol biosynthesis inhibitors			
Triazole	2330	3067	3425
Other azole	795	1192	1465
Other demethylation inhibitor	37	35	39
Morpholine	313	364	415
Other	1399	1683	2300

McDougall, 2016a). Of the 165 active ingredients sold in the agricultural fungicide market, 25 are registered for use on soybeans (Table 9.3). Of these, six account for over half of sales. Many of these compounds are facing intensifying resistance among pathogens, which can limit their effectiveness and scope.

Table 9.3. Soybean fungicides. (Data from Phillips McDougall, 2016a.)

Active ingredient	2015 Sales ($m)	Launch date	Soybean application
Triazoles			
Tebuconazole	530	1988	Foliar, seed treatment
Epoxiconazole	520	1993	Foliar
Cyproconazole	455	1988	Foliar
Propiconazole	310	1980	Foliar
Metconazole	175	1993	Foliar
Flutriafol	90	1984	Foliar, seed treatment
Fluquinconazole	55	1994	Foliar, seed treatment
Tetraconazole	70	1991	Foliar, seed treatment
Other azoles			
Prothioconazole	800	2004	Foliar
Inorganics			
Fentin	75	1954	Foliar
Other multisite			
Fluazinam	135	1988	Foliar
Strobilurins			
Azoxystrobin	1305	1997	Foliar
Pyraclostrobin	850	2002	Foliar
Trifloxystrobin	650	2000	Foliar
Metominostrobin	<30	2000	Foliar
Kresoxim-methyl	130	1996	Foliar
Benzimidazoles			
Thiophanate	195	1968	Foliar
Carbendazim	180	1973	Foliar, seed treatment
Phenylamides			
Metalaxyl	365	1977	Foliar, seed treatment
SDHI			
Fluxapyroxad	390	2012	Foliar, seed treatment
Carboxin	55	1966	Seed treatment
Benzovindiflupyr	230	2013	Foliar, seed treatment
Sedaxane	60	2011	Seed treatment
Others			
Tebufloquin	<10	2013	NA
Ethaboxam	<10	1999	Seed treatment

NA, Information not available.

Chemical Controls for Oomycetes

Of the small number of fungicides registered for use in soybean production, an even smaller number is relevant to the control of oomycetes today, most of which are old chemistries. Control of oomycete diseases relies on products from 16 different product classes, among which the phenylamides, quinone outside inhibitors, carboxylic acid amides, and multisite inhibitors are used most widely. However, resistance has developed against most single-site inhibitors in many oomycete pathogen species, restricting their usefulness. Oomycetes appear to have extraordinary genetic flexibility, which enables them to rapidly adapt to and overcome chemical control measures (e.g. Schmitthenner, 1985; Schettini *et al.*, 1991; Fry and Goodwin, 1997a; Chamnanpunt *et al.*, 2001).

Not only are oomycete pathogens affected in a limited way by many fungicides, but the difficulty of control is magnified because soybean plants are attacked underground, making treatment difficult. The pending phasing out of soil fumigants such as methyl bromide further exacerbates this problem.

When first launched in the 1970s, products with the active ingredient metalaxyl changed expectations for control of oomycete diseases, including those caused by *Pythium* spp. and *Phytophthora sojae*. Metalaxyl rapidly grew in popularity because it offered potent control of all members of the order Peronosporales, including *Pythium*, and allowed flexible application methods, including seed treatment. In 1996, mefenoxam was introduced to the marketplace. Mefenoxam and metalaxyl are stereoisomers, sharing physical properties, efficacy, and known insensitivity issues. They remain the primary phenylamide fungicides relevant to oomycete control in soybeans. More recently, ethaboxam has been commercially introduced as a seed treatment in soybean production.[34] Ethaboxam has been shown to effectively control both *Pythium* spp. and *P. sojae* (Radmer *et al.*, 2017). It is currently being marketed as a seed treatment combined with other fungicides such as metalaxyl or mefenoxam. Hence, the dominant seed treatments used in soybean production today depend on fungicides developed several decades ago.

Use of seed treatments with metalaxyl, mefenoxam, and ethaboxam has followed the growth in conservation tillage in soybean production, which has expanded rapidly with the adoption of herbicide tolerance. Untilled soils remain considerably cooler and wetter for a longer period in the spring, creating favorable conditions for oomycetes, especially *Pythium* spp., which attack slowly emerging seedlings. Treating seed with these fungicides allows farmers not only to protect their crops, but also to plant earlier in the season and optimize the yield potential of their soybeans.

As metalaxyl or mefenoxam is used in practically all seed treatments, a high percentage of production acres rely on the same chemicals for disease control. Resistance has therefore progressed in areas where metalaxyl was used alone, as a curative application, multiple times in a season, and under conditions of high disease pressure (NY DEC, 2015). For instance, in Ohio, *Pythium* spp. resistant to seed treatment products containing metalaxyl have been reported (White *et al.*, 1988; Weiland *et al.*, 2014). Thus, despite its use in seed treatments, sales of the phenylamide fungicide group have declined due to resistance development, and earlier products such as furalaxyl are being discontinued. All the products in the class suffer from some resistance development, with cross-resistance between the compounds.

In order to manage resistance build-up and improve product effectiveness, seed treatments increasingly use multiple chemistries to control oomycete diseases. Pre-mixed products containing combinations of three, four, or more fungicides from multiple chemical classes can have increased effectiveness against other seedling pathogens (e.g. *Rhizoctonia* and *Fusarium* spp.) and utilize multiple modes of action. This approach allows for broader spectrum of activity across the fungal classes known to impact seedling stand establishment, resulting in improved plant health. Seed treatments can also be formulated to include insecticides, nematicides, inoculants, micronutrients and other aids.

Increasing use of seed treatments in soybean production is motivated by a number of factors, first among them being the increased price of soybean seeds. As seed prices have increased with seed functionality (e.g. through the use of biotech traits) and yields, there is a stronger economic motivation for the farmer to protect the seed. Moreover, both early planting and conservation tillage practices have increased risk of seedling diseases, making seed treatments more valuable. Finally, other treatments can help protect yields, for example by stimulating root growth for improved drought tolerance and nutrient uptake, and can reduce the need for other forms of management, such as insecticide sprays.

Battling resistance build-up by combining multiple chemistries and extending their utility is also motivated by the lack of alternatives. We queried several commercial agricultural chemical databases to identify new active ingredients for the control of oomycetes. There are several new fungicides, but few have direct relevance to oomycetes in soybeans. For example, oxathiapiprolin is the first member of a new class of fungicides with activity against oomycete pathogens. It acts via inhibition of a novel fungal target – an oxysterol-binding protein. Oxathiapiprolin facilitates the control of late blight, downy mildews, and *Phytophthora* root and stem blight, but only on fruiting vegetables and cucurbits. Overall, the paucity of new active ingredients for the control of oomycetes in soybean production is striking.

Genetic Disease Resistance: Research and Development

In addition to chemical controls, farmers use genetic resistance to control diseases in soybean production. Desirable features of disease-resistant plants are target specificity, cumulative effectiveness, persistence, harmony with the environment, ease of use, and compatibility with other integrated pest management tactics (Kogan, 1994). Disadvantages include the long development times required to transfer new genetic traits into an established crop species (8–10 years), limits in the sources of genetic resistance, and the ability of pest populations to evolve and overcome host plant resistance. Both public institutions and private firms have long-standing R&D programs in disease control and have used conventional breeding, genetic engineering, and gene editing to confer disease resistance to soybean plants.

Conventional breeding

Conventional breeding transitioned from art to science with the rediscovery of Mendel's work in 1900 (Fisher, 1936). Over the years, conventional breeding has evolved to incorporate new methods and tools and has continued to dominate the development of new crop varieties. Conventional plant breeding relies on the identification of plants with desirable agronomic and phenotypic traits, their combination through genetic crosses, and the selection of progeny with the most desirable combinations of traits. Extensive germplasm collections (e.g. cultivated varieties, old varieties, landraces, wild species) have been developed and screened for sources of host plant resistance to diseases and other useful traits. As the range of potential sources of new traits is limited to the same or closely related species that would produce viable offspring when crossed, broad natural genetic variation in germplasm collections is necessary for success in conventional breeding. Due to this limitation, breeders have used various methods to create additional genetic variation through wide crosses, embryo rescue, tissue culture, somaclonal variation, induced mutagenesis through chemical mutagens or radiation, induced ploidy change, and others.

In its most basic form, breeding is a trait introgression program whereby simple traits that follow Mendel's laws of inheritance are transferred from one crop lineage to another. The line that is receiving the new trait is known as the recurrent parent, while the line providing the trait is called the donor. The offspring of the initial cross between the recurrent parent and donor have a genome that is a combination of the two parent genomes, with half of their genetic material coming from each. The goal of this process is to have a new lineage with a genome as close to identical to the original recurrent parent as possible, with only the new trait added. In order to achieve this, the new progeny are repeatedly backcrossed with the recurrent parent line. Roughly half of the offspring from each cross will carry the new trait and half will not. The breeder must determine which is which and again cross those that carry the new trait with the recurrent parent line. Each time the progeny are backcrossed with the recurrent parent, more of the recurrent parent's genome is recovered; the process is repeated until all or nearly all of the original recurrent parental genome is recovered with the new trait preserved. It takes at least seven backcross generations to recover over 99% of the original recurrent parent genome; depending on the nature and genome location of the new trait, it could take many more.

Before the advent of genetic assay techniques, the genetic composition of parent lines or crossed progeny normally were not known, so individual breeding programs had to grow hundreds or thousands of plant populations and many thousands or millions of individual plants in the field, phenotype them, and harvest them prior to selection (Witcombe and Virk, 2001). Hence, conventional breeding has been an inherently slow and expensive means of genetic improvement. Over the last 10 years, marker-assisted selection (MAS) methods have become commonplace in both private and public breeding programs. Through MAS, unique DNA sequences that are closely linked to genes that govern desirable traits, such as disease resistance, are used to screen each backcross generation and identify individuals that carry the trait of interest prior to flowering. This reduces the number of plants that need to be grown and evaluated in the field, and allows selection and backcrossing to be performed with each generation. Individuals in each new generation also vary in the amount of the recurrent parent genome that is recovered. In some species, markers for the parent genome have been identified, enabling breeders to select not only individuals that carry the trait of interest but also those that recover the greatest proportion of the recurrent parent genome. Thus, MAS has served to dramatically increase the efficiency of conventional crop breeding programs (Collard *et al.*, 2005).

Advancement in the tools and technologies aside, the primary strategy for conferring disease resistance in soybeans has remained the same – identifying sources of host plant resistance in soybean germplasm, introducing the identified resistance into elite soybean lines through several rounds of backcrossing, and doing so as quickly and inexpensively as possible.

Genetic engineering

Conventional breeding is able to exploit only sources of host plant resistance already present in soybean germplasm. Broader opportunities for genetic changes became available, however, when genetic engineering in plants was made possible in the early 1980s. Genetic engineering techniques make it possible to introduce a gene sequence isolated from any living organism into a plant genome; potential sources of new traits are no longer restricted to organisms with which the plant might successfully be crossed. The most well-known use of genetic engineering in crop plants so far has been the development of HT and insect-resistant varieties of major crops such as maize, soybeans, and cotton. In addition to inserting new DNA into a plant, genetic engineering can be used to silence existing (endogenous) genes so they are not expressed in the plant. RNA interference (RNAi) is a natural molecular pathway that all higher organisms use to defend themselves against parasites and pathogens. For uses in genetically engineered plants, non-coding RNA production in the form of double-stranded RNA can be used to set off a chain of molecular events within the cell to silence a gene of interest in the target plant or pest.

Genetic engineering has been used in a variety of ways to impart disease resistance in some plant species. Vincelli (2016) related a range of common strategies, including ways of improving the plant's immune response, disarming plant genes that pathogens use to recognize a potential host, and use of RNAi and other means of silencing pathogen genes that influence pathogen virulence. To date, these methods have only been applied commercially to horticultural crops; no major field crops, including soybeans, contain genetically engineered disease resistance traits (ISAAA, 2018).

Gene editing

One drawback of genetic engineering is that current techniques insert the new DNA sequence into an essentially random location in the target plant genome. Subsequent analysis and breeding to ensure that the new genes are expressed properly and do not interfere with existing plant traits add to the time and expense involved. Gene-editing technologies overcome that limitation by their ability to precisely target a specific DNA sequence for action. Gene-editing methods are still in early stages of development, however, and much of their potential is yet to be realized (van de Wiel *et al.*, 2016). In general, gene-editing techniques use one of a few types of sequence-specific nucleases that bind to a specific DNA sequence and induce a double-strand break (DSB) in the chromosome at that point. They then rely on natural cellular DNA repair mechanisms to complete the editing mission.

Four main classes of sequence-specific nuclease have been used in plant genome editing:

- Meganucleases are naturally occurring molecules present in all types of organism. They naturally bind to and cut DNA chains but are very difficult to customize to a particular purpose. They enjoyed limited adoption and are no longer used to any significant degree.
- Zinc-finger nucleases are a combination of multiple zinc-finger protein domains that bind to DNA and a *Fok*I bacterial nuclease to induce the DSB. They have been used widely but are expensive to customize and produce, and tend to cause significant off-target DSBs.
- Transcription activator-like effector nucleases (TALENs) are structurally similar to zinc-finger nucleases and use the same *Fok*I nuclease. They are less expensive to produce and have proven to be extremely precise.
- The clustered regularly interspaced palindromic repeats (CRISPR) nuclease system consists of a guide RNA construct that binds to a DNA sequence and an endonuclease that induces a DSB. Cas9 nuclease is the most commonly used. Design and synthesis of the guide RNA is quite simple and inexpensive, and little or no off-target effects have been reported in plant applications so far (e.g. Endo *et al.*, 2015; Kim *et al.*, 2017). CRISPR/Cas9 has become the technique of choice for gene editing.

All organisms have two types of DNA repair mechanism that come into play after a sequence-specific nuclease induces a DSB in a gene. In non-homologous end joining, the cell simply repairs the break by sticking the two ends back together. This mechanism is error prone, however, and usually results in the insertion or deletion of one or a few base pairs. This creates a frame-shift mutation that usually causes the gene to completely lose its function. This so-called "knock-out" is by far the most common type of edit employed in crop plants today (Hartung and Schiemann, 2014). Researchers have used knock-out to disable susceptibility genes in some crops, conferring resistance to specific diseases (e.g. Li *et al.*, 2012; Wang *et al.*, 2014).

In homology-directed repair, the cell recreates or adds a DNA sequence at the site of the DSB based on a template present in the cell nucleus. By supplying such a template along with the editing construct, it is possible to insert a new gene conferring a desired trait into the crop genome at a specific location, known as knock-in. This represents an improvement on the performance of previous methods of genetic engineering (Voytas and Gao, 2014). This editing technique has been used to great advantage in animal studies (e.g. Ruan *et al.*, 2015). However, homology-directed repair seems to be less straightforward in plants. So far, there have only been a few examples of successfully introducing new genetic material into crop plants using this technique (Chen and Gao, 2014), and there are no known applications of imparting disease resistance to soybeans or other crops.

Future potential

Overall, conventional breeding, genetic engineering, and gene editing provide a diverse set of tools that can transfer genetic resistance traits to new soybean cultivars. R&D, however, is slow and expensive, particularly in the case of genetic engineering, which is tightly regulated. In fact, increased regulatory costs and delays have been experienced in the development of new biotech traits in all crops. Regulatory costs for the global approval of a new biotech event in the mid-2000s were estimated to be $7.5–15 million (Kalaitzandonakes *et al.*, 2007). Preliminary estimates of regulatory costs for the approval of new biotech events during the 2016–2017 period suggest that such costs have almost doubled (authors' unpublished data and estimates). Other authors estimate regulatory costs for developing a new trait through genetic engineering to be even higher, approximately $35 million, and the remaining R&D costs for bringing the trait to market to be an additional $100 million (Phillips McDougall, 2011). In addition to the monetary cost, the time needed to meet all regulatory requirements is considerable. Firms expect to spend roughly 5 years on compliance issues, more time than any other single phase of product development. In total, firms expect the entire process to take 16.3 years from trait discovery to commercialization and sales (Phillips McDougall, 2011).

It is expected that gene editing could speed up R&D and lower the costs of genetic improvement in plants, including for disease resistance, but the technology is still developing and the regulatory framework remains uncertain (Jones, 2015; Servick, 2015). Ideally, gene editing allows a knock-out or knock-in to be established in a germline in one generation, eliminating the need for the lengthy backcrossing process. However, as editing takes place at the cellular level, at this point developers are only able to edit plant embryos, which must then be regenerated through tissue culture. Not all elite varieties can be regenerated, so in some cases backcrossing is still necessary for trait introgression (Shukla *et al.*, 2009).

As in the case of chemical controls, pest resistance build-up demands ongoing development of novel sources of genetic resistance. As pathogens produce effector molecules to increase virulence, plants respond over time by producing receptors (*R* proteins), which detect the presence or the activity of the pathogen effectors, thereby restoring resistance (Chisholm *et al.*, 2006; Jones and Dangl, 2006). With renewed resistance in the host, pathogens may, over time, evolve to produce new effectors to restore compatibility. In turn, the plant may evolve new *R* proteins. This gene-for-gene "molecular arms race" (Gill *et al.*, 2015) between pathogen effectors and their corresponding *R* proteins has been the source of natural *R* genes used in breeding crops for disease resistance (Jones and Dangl, 2006), but it also indicates their expected gradual depreciation and the need for replacement.

Genetic Resistance Portfolio for Soybean Diseases

All available tools for conferring genetic resistance in soybean cultivars have been actively used in public and private R&D programs. Germplasm improvement through conventional breeding is an ongoing process; disease resistance has been bred into soybeans by this means for decades. Advancement of partial disease resistance in individual cultivars is difficult to trace, as no systematic record of this exists. However, more "discrete" disease resistance traits developed through the use of *Rps* or SCN resistance genes is somewhat more tractable and can provide a picture of the genetic resistance portfolio available through conventional breeding in soybean production.

One way to estimate the prevalence of genetic resistance in soybean cultivars is through yield trial reports. Thousands of yield trials are performed each year across the USA in order to evaluate the performance of new soybean varieties, and these trials provide some information about the traits incorporated in the new commercial soybean germplasm. In Table 9.4, we summarize information from yield trials in different states performed over a 5-year period (2008–2012) and the proportion of new soybean cultivars that contained disease resistance traits. On average, more than 70% of trialed cultivars contained genetic SCN resistance, and some 60% contained one or more of the *Rps* genes conferring resistance to *P. sojae*.

Of course, yield trial data provide information on the proportion of new soybean cultivars with disease resistance traits but not on the share of soybean acres planted with them. Indeed, US soybean farmers in our surveys reported that almost 40% of the soybean seed they purchased contained *Rps* genes (see Chapter 6, this volume). Hence, it appears that soybean genetics suppliers have been incorporating genetic resistance to disease in an increasing share of their varieties through ongoing conventional breeding programs.

In addition to conventional breeding, genetic engineering has also been used to confer disease resistance in soybeans. Permits and notifications required throughout the process of developing biotech disease resistance traits provide a record of such developments over time. As shown in Fig. 9.3, the first such permits were issued more than 25 years ago and their number peaked in 2015. Fungal and nematode resistance has dominated the experimental pipeline, while the number of trials with virus resistance traits has expanded in recent years. Despite the long-standing biotech research activity in the industry and academic institutions, however, biotech disease resistance traits have not

Table 9.4. Disease resistance traits in state yield trials from 2008 to 2012. (Data from state yield trial data, author calculations.)

	New soybean cultivars containing resistance trait (%)	
State	SCN resistance genes	*Rps* genes
Illinois	87	62
Iowa	79	66
Minnesota		64
Missouri	87	46
Indiana		80
Kansas	68	44
Tennessee	71	

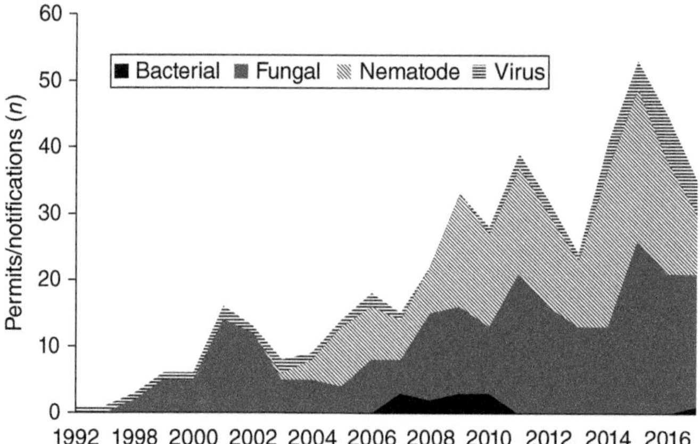

Fig. 9.3. Annual soybean permits and notifications involving disease resistance. (Data from USDA-APHIS Permits, Notifications, and Petitions database at www. aphis.usda.gov/aphis/ourfocus/biotechnology/permits-notifications-petitions, and author calculations.)

reached the commercialization stage. In fact, biotech disease resistance traits appear to be far from commercialization. Once a new biotech trait concept is generated, the potential innovation must progress from proof of concept to full commercialization; there are many technical, regulatory, and commercial hurdles to overcome along the way. We can trace the progress of biotech disease resistance traits in soybeans based on R&D and commercial pipeline reports produced annually by Context Network (Des Moines, Iowa, USA), as well as commercial pipeline reports from publically traded biotech firms. As of 2016, three large biotech firms, BASF, DuPont, and Syngenta, had biotech traits targeting resistance to soybean rust in their development pipeline. Of these, two were still in the initial proof-of-concept stage and one had progressed to stage two, early product development. In addition, in a joint venture, BASF and Monsanto had begun development of an SCN resistance trait, which was in the proof-of-concept stage. Thus, while there has been long-standing activity in biotech R&D for soybean disease resistance, the traits currently in the pipeline are at least a decade away from commercialization, if they prove successful.

Gene-editing methods have demonstrated a great deal of potential for manipulating the genetic make-up of crops, and R&D investments in this area have increased fast. Academic journal articles reporting studies using gene editing have increased at nearly an exponential rate since the discovery of CRISPR/Cas9 (Sternberg and Doudna, 2015). However, much of the contribution made by CRISPR and other gene-editing techniques so far has been in basic gene function research (Liu *et al.*, 2016). A few instances of successfully using gene editing to modify crop plants have been reported, but only two new varieties, HT oilseed rape (Lusser *et al.*, 2012) and high amylopectin (waxy) maize (DuPont Pioneer, 2016), have been commercialized so far.

In the context of crop disease control, researchers have used gene-editing methods to disable susceptibility genes and develop lines of rice resistant to rice blast fungus (Wang *et al.*, 2016) and bacterial blight (Li *et al.*, 2012), as well as wheat resistant to powdery mildew (Wang *et al.*, 2014). These were all cases of knock-outs, point mutations caused by the non-homologous end-joining pathway. Inserting genetic sequences using the homology-directed repair pathway has proven to be much more difficult in plants (van de Wiel *et al.*, 2016). CRISPR has shown some promise in developing soybeans resistant to rust (Langenbach *et al.*, 2016). As partial resistance to oomycete diseases is polygenic and specific resistance is due to *Rps* genes that must be inserted, it is not likely that gene editing will make a significant contribution to soybean disease control until these technical challenges are successfully worked out.

Genetic Resistance Portfolio for Oomycete Control

To better understand the availability of effective genetic resistance to soybean farmers, we once again focus on currently commercialized traits for combatting oomycetes in soybean production, especially *P. sojae*. The disease was first identified in the USA in the late 1940s as soybean production was beginning to take off; however, the causal pathogen was not described until 1958 (Kaufmann and Gerdemann, 1958). Genetic resistance was quickly identified as a potential mechanism for defense against root rot. Two types of resistance have been available: *R* gene-mediated resistance, and partial resistance. In the USA, most soybean cultivars have some level of resistance to *P. sojae*, either through an *Rps* gene or combined with partial resistance. Partial resistance is effective against all races of *P. sojae*, but the level of resistance is not complete. *Rps* genes usually provide absolute protection (i.e. immunity) against target *P. sojae* races. While partial resistance is more stable than single-gene resistance, it is more complicated to effectively introduce into elite germplasm. Accordingly, *Rps* genes have been the focus of much of the effort to mitigate losses to PRR.

Rps genes have been providing reasonable protection against the majority of *P. sojae* populations in the USA for the last four decades. The first *Phytophthora* resistance gene (*Rps*1a) was identified in the 1950s (Bernard *et al.*, 1957). However, it was not until the mid-1970s that other *Rps* genes were identified. *Rps*2 was identified in 1974 (Kilen *et al.*, 1974) with many others being discovered in the late 1970s and early 1980s (Table 9.5). To date, 15 *Rps* genes have been mapped to nine genetic locations including five alleles of *Rps*1 (*Rps*1a, *Rps*1b, *Rps*1c, *Rps*1d, and *Rps*1k) and three alleles of *Rps*3 (*Rps*3a, *Rps*3b, and *Rps*3c) (Sugimoto *et al.*, 2012). Other genetic locations include *Rps*4, *Rps*5, *Rps*6, *Rps*7, and *Rps*8. More recently, a number of novel *Rps* genes/alleles have been reported including *Rps*Yu25 and *Rps*9 (Sun *et al.*, 2011; Wu *et al.*, 2011), and an unnamed *Rps* gene (Sugimoto *et al.*, 2011). Other likely gene/alleles continue to be identified (e.g. Lin *et al.*, 2013).

Only a subset of all *Rps* genes discovered has proved commercially relevant, however. Specifically, seven genes, *Rps*1a, *Rps*1b, *Rps*1c, *Rps*1k, *Rps*3a, *Rps*6, and *Rps*8, have been deployed in commercial soybean varieties. In the 1980s, *Rps*1a and *Rps*1c were the most common *Rps* genes incorporated into

Table 9.5. Initial discovery of *Phytophthora* resistance genes in soybeans. (Adapted from Sugimoto *et al.*, 2012.)

Rps gene	Initial citation
*Rps*1a	Bernard *et al.* (1957)
*Rps*1b	Mueller *et al.* (1978)
*Rps*1c	Mueller *et al.* (1978)
*Rps*1d	Buzzell and Anderson (1992)
*Rps*1k	Bernard and Cremeens (1981)
*Rps*2	Kilen *et al.* (1974)
*Rps*3a	Mueller *et al.* (1978)
*Rps*3b	Ploper *et al.* (1985)
*Rps*3c	–
*Rps*4	Athow *et al.* (1980)
*Rps*5	Buzzell and Anderson (1981)
*Rps*6	Athow and Laviolette (1982)
*Rps*7	Anderson and Buzzell (1992)
*Rps*8	Gordon *et al.* (2006); Sandhu *et al.* (2004)

commercial soybean varieties in the north-central region of the USA (Stewart *et al.*, 2014). When races of the pathogen compatible with these genes became prevalent (Grau *et al.*, 2004), *Rps*1k became the most common resistance gene used (Schmitthenner *et al.*, 1994). *Rps*1k was commercially deployed in 1982 and since that time only one commercially useful new *Rps* gene, *Rps*8, has been discovered; races virulent against it already exist.

Our review of soybean variety yield trials performed in recent years updates and confirms these findings. As we discussed earlier in this chapter, a large number of commercial soybean varieties introduced in the 2008–2012 period have incorporated some resistance genes; we find that these varieties represent well over half of all seed sales in the USA. As expected, *Rps*1c and *Rps*1k appear to be the most popular, with the ageing *Rps*1a only used in a small share of varieties. As an example, Fig. 9.4 shows the utilization of the three most commonly used resistance genes in yield trials in the state of Illinois from 2004 to 2012.

Use of *Rps* genes has created selection pressure. In fact, the number of *P. sojae* races appears to have increased rapidly in recent years (Leitz *et al.*, 2000; Sugimoto *et al.*, 2012). Over 70 different races of *P. sojae* have been detected, with as many as 50 from a single field. Clearly, *P. sojae* is diverse, diversifying, and expanding. This has allowed the adaptation to many of the commercial *Rps* genes that are currently available in soybean cultivars (Nelson *et al.*, 2008). For example, in North Dakota there have been reports of the presence of disease in soybean varieties containing the *Rps*1c and *Rps*1k resistance genes (Nelson *et al.*, 2008). Poor performance of *Phytophthora*-resistant cultivars in Illinois has also been related to a high diversity of field isolates (Malvick and Grunden, 2004), which has also been documented in Iowa (Robertson *et al.*, 2007). Overall, the durability of individual *Rps* gene effectiveness has been estimated at 8–20 years, and many of the commercial *Rps* genes have been in the market much longer than that (Dorrance *et al.*, 2003a; Grau *et al.*, 2004).

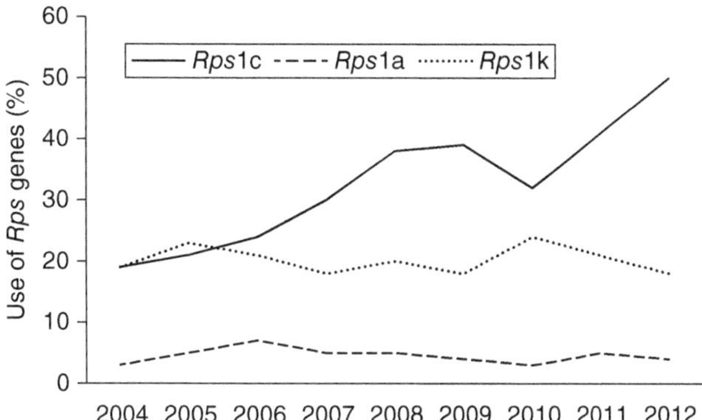

Fig. 9.4. Percentage of marketed soybean varieties containing various *Rps* genes used in the state of Illinois from 2004 to 2012. (Data from Slaminko *et al.*, 2010 (2004–2007 data) and from state yield trial reports and author calculations (2008–2012 data).)

The ageing portfolio of *Rps* genes and the development of pest resistance without immediate replacements are leading genetics suppliers to explore more effective utilization of the existing *Rps* genes. Increasingly, this appears to be through the stacking of *Rps* genes, most notably *Rps*1c/1k and *Rps*3a/1k. Other genes and stacks have also been used, but infrequently. The varieties outside *Rps*1a, *Rps*1c, and *Rps*1k appear to constitute less than 1% of the total number of varieties that are being introduced. Stacking of multiple *Rps* genes is helpful in extending durability, but it is unclear how effective this might prove in the long term. It is worth noting here that there has been limited success in identifying resistance genes against *Pythium* spp., the other relevant oomycetes. Resistance to *Pythium* spp. has been described in only one soybean population, and has not been deployed in any commercial cultivar.

Research to improve the genetic resistance portfolio for the control of oomycetes in soybean production appears to be following several different tracks, including tactical deployment of natural or engineered *R* genes, development of *S* (susceptibility) gene knock-outs, and the use of transgenic hairpin RNA silencing of essential pathogen transcripts (Fawke *et al.*, 2015). Identifying and accurately screening for new *R* genes using molecular markers has proved laborious, expensive, and sometimes problematic due to epistatic interactions between resistance genes. An alternative to marker-assisted screens for identification of novel *R* proteins is effector-based, high-throughput, *in planta* expression assays (Vleeshouwers *et al.*, 2008). Combined with plant disease epidemiology studies and comparative genomics, these expression assays could aid in the prioritization of effectors present in emerging virulent strains as well as those abundant in numerous other isolates (Kamoun *et al.*, 2015). In the last few years, researchers have also begun to adopt structural biology methods to investigate functional relationships between interacting pathogen and plant proteins. Knowledge of how immune receptors function on a molecular level is assisting the development of engineered receptors that detect a broader range of oomycete effectors.

Research to remove key plant genes required for infection is also advancing. These *S* gene mutation-based mechanisms could provide more durable genetic resistance than that of *R* genes because they involve components that are essential for pathogen survival. There are a few *S* genes that show promise as a means to control oomycetes, (e.g. resistance to *Phytophthora palmivora*; Gobbato *et al.*, 2013). Unfortunately, the large majority of *S* genes are involved in essential plant processes and hence knock-outs could result in lethal phenotypes. Therefore, for such *S* genes to be useful agriculturally, different alleles must be identified that encode proteins with reduced but not fully abolished activity. Targeted mutagenesis and assessments of natural variation using haplotype-specific markers could provide solutions. It is worth noting that a number of oomycete genomes have been sequenced to date, including those of *Pythium ultimum*, *Phytophthora infestans*, and *Phytophthora sojae*, (e.g. Tyler *et al.*, 2006). These species could serve as tools to discover more about how oomycetes interact with their hosts and, ultimately, which genes encode effectors, resistance proteins, and susceptibility proteins. Host-induced gene silencing, based on transgenic plants that produce hairpin RNA constructs targeting pathogen transcripts essential for virulence, is also being advanced. The principle has been demonstrated to work in fungi, and accumulating evidence suggests its transferability to *Phytophthora* spp. Overall, these and many other research tracks are promising but are still at early stages of development.

The review of the genetic resistance portfolio for the control of soybean seedling disease and mid-season root rots in this section reveals circumstances similar to those in the portfolio of chemical controls. The technologies available in the market are few and dated, with pest resistance cutting into their overall field effectiveness. Research efforts through conventional breeding, genetic engineering, and gene editing are being advanced to enrich the genetic resistance portfolio, but most of the research pipeline appears to be at early development stages and limited in size.

Determining the Supply of Disease Control Inputs

Our review above indicates that, over the last several decades, and especially in the 1970s and 1980s, the crop protection and seed industries developed and supplied farmers with fungicides and resistant seeds that reduced yield losses from disease. The supply of new chemical controls and genetic resistance traits slowed down, however, in recent years, and the future pipeline of replacement products or products with new modes of action appears limited in size and many years away from commercialization. How are we then to understand these patterns in the supply of disease control inputs for soybean production? For insight, we must look into the economic decision-making process of the suppliers.

Much like farmers, suppliers of chemical controls and resistant seeds decide on whether to invest in the development of such inputs by making choices that maximize their expected profits. In practice, these decisions are quite complex. As we described earlier in this chapter, input suppliers must first make R&D

investments that involve multi-million-dollar allocations for 10–15-year periods before any revenue can be realized. Because of the inherent uncertainty in discovery, not all such investments bear fruit. For those that do, long revenue streams may be available, but in the case of soybean disease control, these streams tend to be limited in size and uncertain.

Because of the diversity of pathogens and environments in soybean production, disease control inputs developed by individual suppliers typically target only specific diseases and have regulatory approvals for only certain jurisdictions. These circumstances limit the overall size of the disease control input market and, hence, the overall revenue stream suppliers may anticipate after commercialization. Furthermore, because of the unpredictable incidence and severity of disease from one geography to another and from one year to the next, such revenues are also quite uncertain at the time of planning. Finally, pest resistance build-up can limit the product life cycle and revenue potential of all such products, so input suppliers must account for such obsolescence in their planning.

Overall, suppliers of chemical controls and genetic resistance must balance long planning horizons, significant upfront costs, and streams of uncertain revenues in making R&D investment decisions for new disease control solutions. For those R&D investments that are made and that prove technically successful, their products must compete for acreage and farmer adoption. A key determinant of farmer adoption that input suppliers do control is the input price. Chemical controls and genetic resistance traits are typically afforded intellectual property protection (e.g. patents) and are differentiated from one another. Under these circumstances, input suppliers are said to have market power (Kalaitzandonakes *et al.*, 2010) and can set the market prices of their products in order to maximize their expected profits. Still, these prices are capped by the FWTP for such inputs, which is based on the economic value of disease control they afford, as described in the two previous chapters. Because soybean farmers are heterogeneous, their WTP for disease control inputs varies widely. As such, if input suppliers set their prices too high, there would be many farmers whose WTP would not be high enough to adopt and use such inputs. Suppliers can encourage adoption by lowering their prices.

We can illustrate these considerations of input suppliers with an aggregate demand curve that we constructed from the WTP of US soybean farmers for a broad genetic resistance trait against seedling disease, as discussed in the previous chapter and illustrated in Fig. 8.8. The overall market size for such a trait is limited by the number of acres that do not experience seedling disease and mid-season rots. Farmers responding to our survey reported that approximately 64% of soybean acreage had no recorded occurrence of seedling disease over the preceding 10 years. For this market segment, the WTP of farmers is naturally very low. Given these circumstances, as Fig. 8.8 shows, if the price of such a trait was set by the suppliers at $7 per bag of seed, it would be adopted on about 30% of soybean acres. Raising the price to $14 per bag would reduce adoption to only 10% of soybean acres, a rather steep decline. Similar pricing scenarios demonstrate that trait suppliers are forced to balance prices with adoption levels.

There are other checks on supplier prices and profitability. As we illustrated in Chapter 7 (this volume), lower soybean prices limit the economic value of disease control on soybean farms. Under such conditions, the demand function shifts to the left, and input suppliers must reduce their prices to preserve farmer adoption. Input suppliers must also compete for farmer adoption with other disease control inputs. It is important to emphasize here that such competition is not simply between like forms (e.g. one genetic resistance trait against another). Genetic resistance competes for acres with seed treatments and other methods. As we described in Chapters 7 and 8 (this volume), changes in the relative effectiveness or the relative prices of different disease control practices shift their relative contribution to farm profitability, their levels of adoption, and the overall demand by soybean farmers. Hence, while suppliers of disease control inputs are able to vary their prices to maximize their profits, they are constrained by the economic value that their products generate on the farm relative to all other alternatives.

Given all these factors, at the planning stage suppliers anticipate long streams of revenues from disease control products they can develop; however, these are constrained by the expected relative effectiveness and unit costs of other disease control practices available, the levels of soybean prices, the incidence and severity of disease, and other factors. The duration of these revenue streams are conditioned by the rate of technical obsolescence of individual products (e.g. the emergence of better-competing disease control practices, pest resistance build-up, or increased regulatory restrictions). Overall, the discounted stream of expected variable profits (revenues less manufacturing, distribution, and operating costs) for any given disease control product considered by input suppliers must comfortably exceed the expected R&D costs to warrant further consideration for a position in the supplier's R&D portfolio.

Positive expected profitability is necessary but not sufficient for the decision to invest in the R&D of a disease control product. Suppliers have limited R&D budgets and choose to invest in only the subset of products with the highest expected profitability. In this respect, R&D investments in soybean disease control inputs are often handicapped against alternatives with larger market potential and more predictable revenue streams. For instance, investing in the R&D of a broad-spectrum herbicide for soybean production may be deemed preferable by suppliers, as weeds are omnipresent in all geographies and in all years, and, hence, in all soybean acres.

These circumstances, along with intervening conditions of increasing regulation, rising pest resistance, and declining "hit" rates in research and discovery, may explain the slowdown we have observed in the development of new disease control inputs for soybean production in recent years. They may also explain the prevailing strategy of stacking existing molecules and genes in order to extend their usefulness at a more modest R&D cost.

Summary

Chemical controls and genetic resistance are the primary means used by soybean farmers to fight disease. R&D for new disease control inputs is a long, slow, costly, and uncertain process. As a result, disease control input needs identified today may yield commercial inputs that farmers can use many years in the future. In the last several decades, the crop protection and seed industries have increased their R&D spending and have capitalized on accelerating innovation in tools and methods. Even under such conditions, however, the pace of new product introduction appears to be slowing in both of these input industries. The supply of new disease control inputs for soybean production has certainly experienced such declining trends. Whatever the causes, the end result is that innovation appears to be slowing down at a time when pathogen evolution is presenting producers with increasing problems of resistance to existing controls. Over the short term, developers are managing pest resistance by combining existing chemical molecules and stacking existing disease resistance traits. The longer-term effectiveness of such practices is still unclear.

The supply of new inputs for disease control to complement and replace an ageing portfolio will be essential in the future. Such a supply is constrained by the underlying economics of soybean disease control. Unpredictable and infrequent disease incidence and severity limit the economic value of control practices and FWTP for them. The broad diversity of diseases affecting soybean acres also fragments the disease control input market and caps the overall market potential for any single disease control input.

Input suppliers in the crop protection and seed industries also operate in an environment of uncertainty related to the timing and success of discovery, the regulatory environment, and the likely durability of their commercial products. Thus, uncertainty on both sides of the disease management process can drive a wedge between what producers are willing to pay and what developers believe they need to receive to fund continuing R&D. As a result, there may be chronic underinvestment in new agricultural disease management technologies.

Note

[34] Ethaboxam was registered for use as a soybean seed treatment in 2015. However, it is not a new fungicide. It was discovered in 1993 and first registered for use in South Korea in 1999. Over the years, it has been used for foliar applications in corn, fruits, and vegetables.

10 Economic Benefits from Innovation

Most soybean farmers use chemical fungicides and resistant cultivars to limit yield losses from disease when they expect damage from pathogens to exceed the EIL. The suppliers of such inputs actively invest in R&D in order to regularly improve them by developing more efficient genetic resistance traits or novel active ingredients and improved formulations. In this chapter, we examine who benefits from such innovations and to what extent. We generally expect that soybean farmers who adopt improved disease control inputs will see an economic benefit. As long as their expectations about the potential damage from disease and the relative profitability of different control practices are sufficiently close to actual conditions, they should realize higher profits when they decide to use these innovations in their fields. We also expect the input suppliers that develop the new resistant seeds and pesticides to benefit. As long as their expectations about their R&D costs, farmer demand, and resistance build-up are within range, they should realize higher profits from the development of innovations. As we discuss in this chapter, however, consumers are the group that benefits the most from such innovations. When soybean supplies expand due to improved disease control, soybean prices decrease; as a result, food products derived from soybeans become more plentiful and less expensive.

To illustrate how large the benefits from innovations in soybean disease control may be, as well as how such benefits may be distributed among various stakeholder groups, we once again focus our empirical analysis on a single innovation – a conjectural genetic trait offering broad resistance against seedling disease and mid-season root rots. Our analysis begins with the farmer's decision to adopt the innovation. This is the same decision we examined in Chapter 8 (this volume), but here we abstract from farmers' expectations on disease incidence and severity, the relative effectiveness of different disease control methods, prices, costs, and other such considerations. Instead, our starting point is the farmer's decision to opt for the new seeds with the broad disease resistance trait rather than seeds with fungicide treatments, partial disease resistance, *Rps* genes, or no disease control at all. In this context, we measure the change in farm profitability when the novel broad resistance trait is adopted to replace an existing disease control practice or input. A positive change in farm profitability indicates an economic benefit for the farmer who adopts the innovation.

Summing up the economic benefits derived from the novel seeds across all adopters provides an aggregate farm-level measure of the innovation benefits. However, such a measure represents only the "initial incidence" of benefits from adoption of the innovation, rather than the "final incidence," which accounts for the shifting of the benefits from farmers to consumers and among different types of farmers as soybean prices are induced to change. As such, we develop and implement methods that account for the redistribution of the farm-level benefits. Thus, in this chapter, we estimate the potential global benefits from adoption of the broad resistance trait and the potential distribution of these benefits among technology suppliers, farmers, and consumers in a large number of countries, only some of which are assumed to adopt the innovation.

Our analysis is forward looking, in the sense that we estimate the potential benefits from the adoption of the innovation over a 10-year period from its initial introduction, assumed here to be from 2018 to 2027. The size of the innovation-induced shift in aggregate soybean supply depends on the impact of the innovation on yields and costs. We use information from our farmer surveys to derive estimates of the potential farm-level consequences of adoption, like those presented in Chapter 8 (this volume). We combine the estimates of farm-level impacts with information on the potential rate of adoption, also derived from our farmer surveys, to parameterize the implied shifts in market supply in the context of a global multi-market simulation model. The model produces simulated market outcomes in terms of prices and quantities of farm commodities, both with and without the adoption of soybean varieties with the broad resistance trait. It includes a full set of interactions between soybeans and complement and substitute crops, and the effects of international trade. We then use these estimates as inputs to calculate the net impacts on the welfare of innovators, adopting and non-adopting farmers, and consumers worldwide. We begin by presenting the conceptual models we have used to estimate farm-level and global benefits of innovation. We subsequently implement these models empirically and discuss the results.

Benefits of Innovation: Conceptual Foundation

Farm-level impacts and adoption decisions

The introduction of broad resistance to seedling disease and mid-season root rots would give soybean farmers one more option for disease control. The new resistance trait could increase yields through more effective disease management and could lower production costs by reducing expenditures on other inputs. For example, fungicide use could be reduced or eliminated, and agronomic practices such as reduced tillage could be implemented more widely, leading to a reduction in the variable costs associated with machinery use, fuel, and labor. The potential economic advantages of the new cultivars would provide incentives for farmers to consider adoption. Farmers would choose their optimal, profit-maximizing input mix in view of the relative prices of conventional and resistant seeds, fungicides, and other inputs.

An individual farmer's decision to adopt the new cultivars can be, as before, represented with a simple model. Following the damage abatement model in Chapter 4 (this volume), the profit-maximizing input allocation decision of the price-taking farmer can be represented as:

$$\max_{z,x} \Pi = pg(Z)\left[1 - D(N,X)\right] - K(Z) - C(X) \tag{10.1a}$$

where Z is a vector of agronomic inputs (e.g. fertilizer, capital, labor) and X is a vector of disease control inputs (e.g. pesticides, resistant seeds), while $K(Z)$ and $C(X)$ are their cost functions, respectively. For a given disease population and disease control management, we can replace the damage function $g(Z)[1 - D(N, X)]$ with the expected yield Y and restate the farmer's decision as:

$$\max_{z,x} \Pi = pY(Z,X) - K(Z) - C(X) \tag{10.1b}$$

As we saw in Chapter 8 (this volume), an individual farmer will decide to adopt the new cultivars with the broad resistance trait by comparing their expected incremental improvements in farm profit against existing practices. We can say, then, that farmer i will adopt a resistant soybean cultivar j in year t if it is expected to be more profitable than the next best alternative k under the following conditions:

$$a_{ijt} = \begin{cases} 1 \text{ if } \pi_{ijt} > \pi_{ikt} \\ 0 \text{ if } \pi_{ijt} < \pi_{ikt} \end{cases} \tag{10.2}$$

where $\pi_{it} = P_{it}Y_{it} - VC_{it} - S_{it} - F_{it}$.

Here, a_{ijt} is an indicator variable that is equal to 1 if farmer i adopts cultivar j in year t and zero otherwise, while π_{it} is farm profits, P_{it} is the price per bushel of soybeans in year t, Y_{it} is the yield in bushels/acre, VC_{it} is variable cost of production other than seed and fungicide costs, S_{it} is seed cost per acre, and F_{it} is the cost of fungicide, used either as seed treatment or spray application. Eqn 10.2 suggests that the farmer will adopt the novel cultivar when it is more profitable than the best existing alternative. In the unlikely event that $\pi_{ijt} = \pi_{ikt}$, the farmer will be indifferent between the two options.

The benefit per acre on those acres using the novel resistance trait is simply $B_{it} = \pi_{ikt} - \pi_{ijt} = \Delta\pi_{it}$; based on Eqn 10.2 it can be expressed as:

$$\begin{aligned} B_{it} &= \left(\frac{\Delta Y_{it}}{Y_{it}} - \frac{\Delta VC_{it}}{P_{it}Y_{it}} - \frac{\Delta S_{it}}{P_{it}Y_{it}} - \frac{\Delta F_{it}}{P_{it}Y_{it}} \right) P_{it}Y_{it} \\ &= \left(y_{it} - \theta_v v_{it} - \theta_s s_{it} - \theta_f f_{it} \right) P_{it}Y_{it} \end{aligned} \tag{10.3}$$

where y is the proportional change in yield per acre, v is the proportional change in the costs per acre of agronomic inputs, θ_v is these agronomic input costs as a share of total revenue per acre, s is the proportional change in seed costs per acre, θ_s is seed costs as a share of total revenue per acre, f is the proportional change in fungicide costs, and θ_f is fungicide costs as a share of total revenue per acre. Hence, B_{it} represents the economic benefit per acre from the introduction of the novel genetic resistant trait. As such, the total annual benefits from adoption for a particular farmer i in a particular year t, FB_{it}, can be obtained by multiplying the per-acre benefit by the number of acres planted to the new variety by that farmer, A_{it}:

$$FB_{it} = \left(y_{it} - \theta_v v_{it} - \theta_s s_{it} - \theta_f f_{it}\right) P_{it} Y_{it} A_{it} \tag{10.4}$$

Estimating the farm-level production and economic impacts of improved disease management is only the first step, however. We are also interested in how these farm-level changes in costs and yield influence production, consumption, trade, and social welfare in the wider economy.

Aggregate measures of benefits when prices are unaffected

If we have data on the elements of Eqn 10.4 for the adopting farmers and corresponding data on the adoption rate and the gross value of production, we can obtain a reasonable measure of aggregate gross annual farmer benefits of adoption, TFB_t, defined as the portion of the gross value of production resulting from the change in productivity and cost-efficiency associated with improved disease management:

$$TFB_t = \sum_{i=1}^{n} FB_{it} = \sum_{i=1}^{n} \left(y_{it} - \theta_v v_{it} - \theta_s s_{it} - \theta_f f_{it}\right) P_{it} Y_{it} A_{it} \tag{10.5}$$

By defining K_{it} as:

$$K_{it} = y_{it} - \theta_v v_{it} - \theta_s s_{it} - \theta_f f_{it} \tag{10.6}$$

it is clear that the proportional change in the gross value of production from the innovation, K_{it}, is the weighted sum of the proportional yield and cost shifts induced by its adoption.[35]

The measure in Eqn 10.5 does not include the annual benefit to the technology developers who would be the source of the genetic resistance trait, TSB_t. This benefit could be measured in terms of the element of Eqn 10.5 associated with the seed premium paid by the farmer:

$$TSB_t = \sum_{i=1}^{n} \theta_s s_{it} P_{it} Y_{it} A_{it} \tag{10.7}$$

This is a measure of the additional profits that seed firms would derive from introduction of the innovation, given the price of seed, the assumed pattern of adoption, and the effect on yield. It is a gross rather than a net benefit in that it does not include the additional expenses incurred in developing and marketing the novel disease-resistant seeds. It also does not include the loss of revenue from lost sales in seeds with older resistance traits and replaced fungicides.

Combining Eqns 10.5 and 10.7 provides a measure of total gross annual benefits to both farmers and technology providers from adoption of the novel disease resistance trait, TB_t, defined as:

$$TB_t = TFB_t + TSB_t = \sum_{i=1}^{n} \left(y_{it} - \theta_v v_{it} - \theta_f f_{it} \right) P_{it} Y_{it} A_{it} \tag{10.8}$$

The above results are derived under the assumption that adoption of the innovation does not significantly change the total quantity of soybeans produced, and thus has no effect on the price of soybeans. Of course, with sufficiently high levels of adoption, aggregate soybean supplies can increase enough to push soybean prices lower. In addition, changes in the soybean market could have further effects on prices and quantities in the markets for substitute and complement commodities. All such price and quantity changes can affect consumer and producer welfare. Accounting for these market effects may not significantly change the total level of benefits from the adoption of the innovation, but the distribution of benefits is likely to be substantially different.

Market Effects and the Distribution of Benefits

Modeling and measuring the aggregate economic impacts of technological change typically involves a conventional supply and demand framework in which innovations are represented as shifts in supply. This approach, which is discussed in detail by Alston *et al.* (1995, pp. 207–221), can be applied to the potential adoption of the novel broad resistance trait with yield improvements and cost-efficiency gains being the source of the supply shift, as measured by K_{it} above.

Standard single-commodity market model

In the standard single-commodity market model, shown in Fig. 10.1, adoption of a technological innovation increases productivity, in terms of either greater production or reduced costs, or both. The market outcome is the same in all cases: the commodity supply curve shifts down/out against a stationary demand curve, bringing about an increase in quantity produced and consumed, and a decrease in price. The benefits are assessed using so-called Marshallian measures of consumer and producer surplus. The size of the benefits from the innovation depends primarily on the magnitude of the shift in the supply curve. Innovations can have other, second-order effects on total benefits, but these are typically limited in scope. The slopes of the demand and supply curves have important implications for distribution of the benefits between farmers and consumers.

We can use the standard single-commodity model to illustrate the basic market outcomes from innovation and how those outcomes are affected by

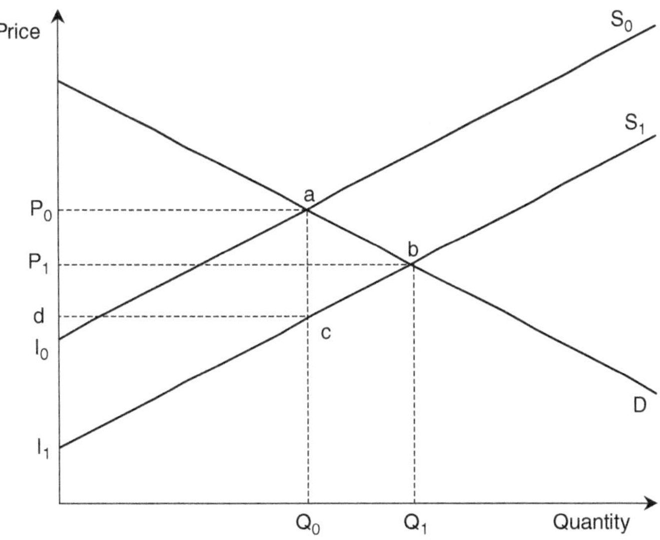

Fig. 10.1. Standard single-commodity market model. 0, base line; 1, the innovation scenario; D, demand; I, intercept; P, price; Q, quantity; S, supply.

the nature of supply and demand relationships. In Fig. 10.1, the increase in production quantity from the novel resistance trait is illustrated by the downward shift in the supply curve from S_0 to S_1, resulting in a greater supply at any given market price. Due to the nature of demand, the greater supply can only be sold at a lower price. Thus the market will see an increase in quantity from Q_0 to Q_1 and a decrease in price from P_0 to P_1. The total annual market benefit from innovation is the sum of all the additional market transactions that take place because of the changes in price and quantity. The amount of benefit can be visualized as the area under the demand curve and between the two supply curves, the trapezoid I_0abI_1.

The distribution of the benefits between farmers and consumers depends on the relative elasticities of supply and demand, expressed as the slopes of the respective curves, the nature of the shift in supply and, less importantly, the functional forms of supply and demand. The benefit to consumers of greater quantities at lower prices is obvious, and the magnitude of the benefit can be visualized as the area P_0abP_1. The benefit to farmers is less straightforward. It is represented by the difference between the producer surplus in the presence of the innovation, the area P_1bI_1, and the producer surplus in the absence of the innovation, the area P_0aI_0.

Here, we must note that the model depicted in Fig. 10.1 is a special case on two counts. First, the nature of the technology-induced supply shift is a critical choice because it has a strong influence on the distribution of benefits; it is also not easy to observe. In this example, we assume a vertically parallel supply shift in order to maintain simplicity. The benefit to farmers, the difference between the two triangular areas described above, then comes out to be equal to the area P_1bcd, the reflection of the consumers' benefit. If the elasticity (slope) of the supply curve were to change along with the shift, this would not be the case. Second, the elasticities of the supply and demand curves are the inverse of one another, i.e. the angles at which the two curves intersect the vertical axis are the same. In this special case, farmers and consumers share the market benefit equally. As the demand curve becomes less elastic (steeper) than the supply curve, more of the benefit is shifted to consumers. Likewise, if the supply curve is steeper, more of the benefit is allocated to farmers. Finally, we can see that, in the case of a positive supply shift, as depicted here, consumers will always benefit to some extent. Farmers, however, may experience benefits or losses, depending on the relative elasticities of supply and demand and the nature of the supply curve shift.

The single-commodity model is useful in illustrating the various market forces that influence the size and distribution of benefits from innovation, even though it is not a strictly accurate representation of real markets. In order to represent world commodity markets realistically, we must expand the model to include multiple national and regional markets and multiple agricultural commodities. Different national and regional markets must enter the model separately, as they may experience differing prices due to different tariff policies or transportation costs. Markets for complementary and substitute grains and oilseeds, which have their own unique supply and demand conditions, will also be affected and must be modeled explicitly (e.g. Freebairn, 1992; Frisvold, 1997; Wohlgenant, 1997; Zhao *et al.*, 2000). To illustrate the redistribution of innovation benefits in the context of interacting markets, we use a simple two-country model. We then present a more realistic multi-country, multi-commodity model, which we use to estimate the benefits of innovation and their distribution in our empirical application.

Standard two-country model

The standard two-country market model, shown in Fig. 10.2, illustrates our general approach. The basic model presents a simplified case in which the novel disease-resistant soybean cultivars are commercialized in a large exporting economy (region A). The rest of the world (region B) sees the results of the innovation through trade but does not adopt it in this model. The model estimates the impact of the innovation on the volume of trade between the two regions and on commodity prices in the two regions.

Figure 10.2A–C represents, respectively, supply and demand in the innovating region (region A), the interaction of excess supply and excess demand in world trade, and supply and demand in the rest of the world (region B). For ease of illustration, all of the supply and demand curves are assumed to be approximately linear. The intersection of the initial excess supply (ES_0) and excess demand curves determines the initial price on the world market, P_0. When the new disease-resistant soybeans are introduced in country A, the increased yield produces a parallel shift in domestic supply from S_{A0} to S_{A1}, as was depicted in the single-market model in Fig. 10.1, and in consequence the excess supply on world markets shifts from ES_0 to ES_1 and the world price falls from P_0 to P_1.

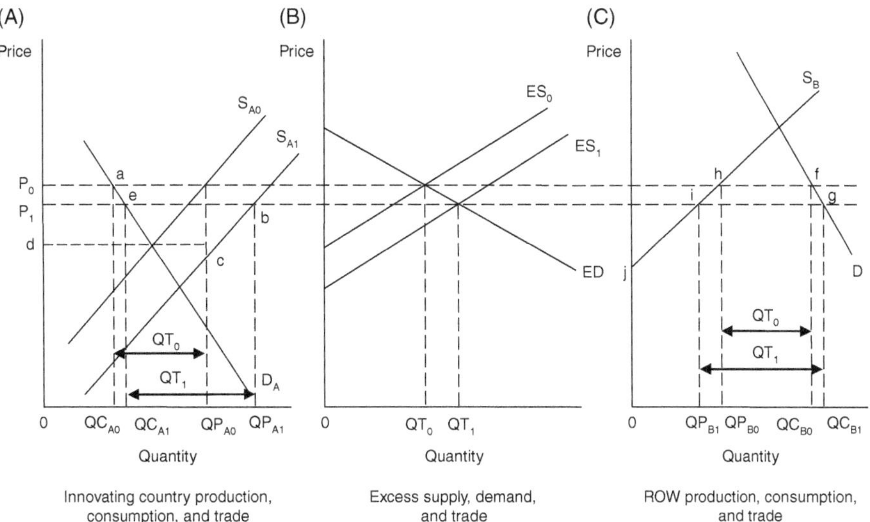

Fig. 10.2. Standard two-country model of impacts of innovation. 0, base line; 1, the innovation scenario; region A, innovating country A; region B, international trade; C, quantity consumed; D, demand; ED, excess demand; ES, excess supply; P, price; Q, quantity produced; QT, quantity traded; region C, ROW, rest of the world; S, supply. (Adapted from Alston *et al.*, 1995.)

Consumers worldwide unambiguously benefit. The benefits to consumers in country A are shown in the figure as the area P_0aeP_1 in Fig. 10.2a and the benefits to consumers in the rest of the world are shown as the area P_0fgP_1 in Fig. 10.2c. Farmers in region A benefit as long as the price reduction from P_0 to P_1 is smaller than the initial vertical supply shift in country A. The benefits to farmers in region A are given by the area P_1bcd in Fig. 10.2a; if world market conditions were such that the new price, P_1, were below point d, farmers in region A would see a net loss. Farmers in region B are net losers, as they are faced with a lower world commodity price without the offsetting cost reductions and yield increases from the innovation. This condition is reflected in the figure by the fact that the area P_1ij is smaller than the area P_0hj.

The above two-country model provides the conceptual foundation for the empirical model we used in this study to estimate the potential global benefits from the introduction of the novel broad resistance trait against seedling disease and mid-season root rots. We present the details of the model we used for our empirical analysis next.

Empirical Models of Economic Benefits from Innovation

To estimate the potential global benefits of the conjectural novel broad genetic resistance against seedling disease and mid-season root rot in soybean production over the period 2018–2027, we develop a detailed partial equilibrium simulation model that captures the linkages among oilseeds and competing crops and the rest of the agricultural sector in markets around the world. For instance, the relationship between demand for soy meal or other protein meals and the livestock sector is modeled explicitly. The model allows estimation of price effects, land reallocation patterns, and substitution among oilseeds and between oilseeds and other crops as a result of the changes in production volume and price. The model is calibrated such that the baseline simulation replicates the actual values of supply, demand, acreage, stocks, and other variables of interest for soybeans and all other major agricultural commodities over a period ending in 2017. For the scenario analysis in the 2018–2027 period, this baseline is simply extended into the future according to their historical values; normal year-to-year variation in prices, production, and so forth are defined by prevailing trends.

The "innovation scenario" describing a world where soybean cultivars with the novel broad resistance trait are available is simulated by incorporating the anticipated changes in yields, costs, and profits caused by introduction of the innovation. The resulting innovation scenario prices, quantities, acreages, stocks, imports, exports, and other indicators are then compared with the baseline values to account for the effects of the innovation. The differences between the innovation scenario and baseline are used to estimate the benefits from the innovation, in the form of changes in consumer and producer surplus, which are defined below.

Model structure

The model used here includes a set of supply and demand equations for each commodity of interest. Separate supply and demand functions are specified for the various oilseeds, their derivative oils and meals, and substitute and complement commodities for each of the 47 countries/regions represented in the model, including Argentina, Bolivia, Brazil, Canada, China, the EU, India, Indonesia, Japan, Malaysia, Mexico, Paraguay, Russia, South Africa, South Korea, Uruguay, and the USA. The following equations, which characterize the market clearing conditions for a particular oilseed commodity and its oil and meal products in a given country, illustrate the general structure of the model:

$$BSt_t = ESt_{t-1} \qquad \text{(oilseeds, meals, and oils)}$$
$$QP = A \times Y \qquad \text{(oilseeds)}$$
$$QP = Cr \times CrY \qquad \text{(meals and oils)}$$
$$TS = BSt + QP + I \qquad \text{(oilseeds, meals, and oils)}$$
$$TD = Cr + Food + Other + X + ESt \qquad \text{(oilseeds)}$$
$$TD = Food + Feed + Ind + X + ESt \qquad \text{(meals and oils)}$$
$$QC = Cr + Food + Other + ESt \qquad \text{(oilseeds)}$$
$$QC = Food + Feed + Ind + ESt \qquad \text{(meals and oils)}$$

Here, BSt refers to beginning stocks of each commodity and ESt refers to ending stocks. When used, the subscript t denotes the current year and $t - 1$

denotes the previous year. *QP* is quantity produced, *A* is harvested acreage, and *Y* is yield in bushels/acre. *Cr* is the quantity of soybeans crushed for meal and oil, while *CrY* is the yield, in percentage terms, of meal and oil, respectively, compared with the original quantity before crushing. *TS* is total supply on world markets and *TD* is total world demand. *I* refers to import quantity for each country and *X* to exports. *QC* is the quantity consumed in each country. The various categories of use are *Food* (humans), *Feed* (livestock), *Ind* (industrial), and *Other*.

Similar supply and demand relationships are included for all other major crop commodities (e.g. maize, wheat, rice, cotton), as well as livestock commodities (e.g. poultry, beef, pork). Within the model, the country-specific prices of each commodity are linked to those of every other trading country using price linkage equations that include import tariffs, taxes, transport costs, and other relevant shifters. Changes in yields and costs from the adoption of the innovation are incorporated directly into the soybean yield and cost functions of the adopting countries.

In order to calculate the change in producer and consumer surpluses caused by the innovation, we follow Alston *et al.* (1995) and use the following formulae:

$$\Delta PS_{RS} = P_0 Q_{R0} \left(K - Z \right) \left(1 + 0.5 Z \varepsilon_S \right) \tag{10.9a}$$

$$\Delta PS_{RO} = -P_0 Q_{R0} Z \left(1 + 0.5 Z \varepsilon_O \right) \tag{10.9b}$$

$$\Delta CS_R = \left(P_0 - P_1 \right) C_{R0} + 0.5 \left(C_{R1} - C_{R0} \right) \left(P_0 - P_1 \right) \tag{10.9c}$$

where ΔPS is the change in producer surplus; ΔCS is the change in consumer surplus; R denotes a country or region of interest; S denotes soybeans; O denotes other crops including sunflower, rapeseed, and palm oil; P_0 is the baseline price; P_1 is the price in the innovation scenario; Q_{R0} denotes a baseline quantity produced; C_{R0} is a baseline quantity consumed; C_{R1} denotes an innovation scenario quantity consumed; ε_S is the elasticity of supply of soybeans; ε_O denotes elasticities of supply of other crops; K is the percentage vertical shift in the supply function of soybeans resulting from introduction of the novel broad disease resistance trait measured as described in Eqn 10.6; and Z is the relative price change given by $-(P_1 - P_0)/P_0$.

Model calibration and assumptions

An important task in the scenario development is setting the values for key parameters, including the potential impacts of the new broad resistance trait on soybean yields and on production costs, as well as its level of adoption. We therefore begin our scenario development here by setting specific values for these and other relevant parameters. To do so, we use figures from our farmer and CCA surveys as well as the analysis in previous chapters.

The baseline scenario: current conditions and practices

The baseline run of our empirical model reflects the current practices of soybean farmers. Many US soybean farmers currently control seedling disease and mid-season root rots through fungicides and cultivars with partial resistance and *Rps* genes. Specifically, they use seed treatments on some 60% of US soybean acres, while they plant cultivars with *Rps* genes and partial resistance on over 40% of them (see Chapter 6, this volume). The cost of planting cultivars with *Rps* genes is $4.10/acre, while seed treatments cost $7.10/acre, as identified from our farmer surveys (see Chapter 7, this volume). Soybean farmers using both inputs would therefore incur a cost of $11.20/acre.

We estimated previously that the resistance (i.e. *Rps*) genes reduce the yield loss from seedling disease by an average of 24% and from PRR by 51%. Seed treatments reduce the yield loss from seedling disease by an average of 59% and are ineffective against PRR (see Chapter 7, this volume). As these inputs provide only partial control, in any given year, soybean farmers with prior occurrence of seedling disease in their fields face a 24% chance they will experience seedling disease (2.4 years out of 10) with an associated yield loss of 8.2 bushels/acre. Similarly, soybean farmers with prior occurrence of PRR face a 22.7% chance (2.27 years out of 10) of incurring a 7.9 bushels/acre yield loss to the disease (see Chapter 5, this volume).

The innovation scenario

For our innovation scenario, we assume that the novel trait is introduced only in the USA in 2018 and we analyze its potential global economic benefits and their distribution among stakeholders over the first 10 years of its use, in the 2018–2027 period. As with all agricultural innovations, adoption of the novel resistance trait is expected to occur gradually, as farmers would initially experiment on a limited number of acres to assess its relative performance. As their own experience and that of other farmers increases, adoption would expand.

The adoption path would also be influenced by the relative pricing of the novel trait. Given the differential WTP of soybean farmers, the market price of the trait would influence the rate and level of its adoption. Here, we use the demand function we derived in Chapter 8 (this volume) to project the level of adoption of the broad resistance trait. We assume that the suppliers of the innovation would price the novel trait at $5/acre and, as such, adoption would top out at 60% of all US soybeans acres (see Fig. 8.8). We assume that this level of adoption is achieved over the 10-year period of our analysis and that the adoption path follows a typical sigmoidal pattern, similar to those observed in the adoption of insect resistance and herbicide tolerance biotech traits in US soybean and maize production (Fig. 10.3).

Given these conditions and based on our analysis in Chapter 8 (this volume), we assume here that the novel resistance trait would replace fungicide treatments, *Rps* genes, and partial resistance in the control of seedling disease and mid-season root rots. Hence, for 60% of the soybean farmers who use fungicide seed treatments, the substitution would imply savings of $7.10/acre. Similarly, for almost 40% of the farmers who use *Rps* genes and partial resistance, the

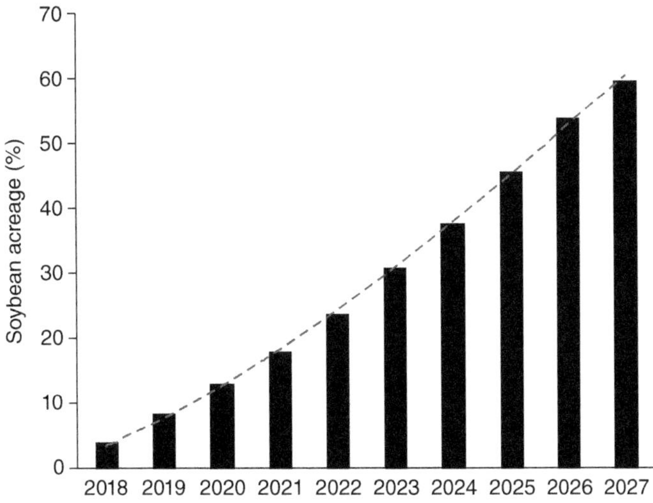

Fig. 10.3. Adoption path of the broad resistance trait in the USA. (Data from author models.)

substitution would result in savings of $4.10/acre. For the farmers who use both practices, the savings would be $11.20/acre.

Soybean yields would also improve for farmers with prior occurrence of seedling disease and mid-season root rots in their fields. As in Chapter 8 (this volume), we assume that disease control with the novel broad resistance trait would improve relative to currently used practices, as the new trait is assumed to reduce yield losses from the disease by 90%. Under these conditions, for the one-third of all US soybean acres with seedling disease, average yields would improve by 1.77 bushels/acre, while for the soybean acres with mid-season rots, average yields would improve by 1.59 bushels/acre.[36]

We also assumed that more effective seedling disease control would encourage a further shift from conventional to no-till and minimum tillage practices. Our CCA surveys indicated that farmers might decrease the use of conventional tillage by as much as 14% in the aggregate if oomycetes were managed effectively. If reduced or no tillage was to be used on these acres, farmers would experience reduced costs and potentially higher yields. Our farmer surveys indicated a net benefit from no-till of $27/acre and $14/acre for reduced tillage (see Chapter 6, this volume). In our simulation model, we applied the mid-point of such cost savings on 14% of the conventional tillage acreage that would adopt the novel resistance trait and we assumed no yield gains from such conversions. The CCAs also estimated that, on average, 29% of farmers would also plant their soybeans earlier in the season if seedling disease was controlled more effectively (Table 6.5). Earlier planting could increase average yields, but because the number of acres that would be affected is unclear, we do not consider such possible yield effects here.

Soybean farmers are heterogeneous and some would benefit more than others from the adoption of the new broad resistance trait (see Chapter 8, this volume). Those who expect greater economic benefits would tend to be early adopters. In such cases, the benefits per acre would tend to be higher in the early years of adoption than in later ones. Nevertheless, because there are other factors that influence adoption (e.g. risk perceptions and preferences, quality of information) we assume here that the per-acre yield and cost-efficiency gains from adoption are even across all years of adoption.

Size of Economic Impact and Distribution of Benefits

Under the conditions set out in our innovation scenario above, the USA would be the only country with direct benefits from the novel resistance trait as it is assumed to be the only region where the new technology would be available for use. Nevertheless, there are significant indirect benefits from adoption of the novel trait that are determined from the market effects that follow its adoption. Given the anticipated adoption path and efficacy of the novel trait, the yield and cost-efficiency gains would induce the US soybean supply to expand. Increases in soybean production in the USA would be almost 85 million bushels above baseline at full adoption, in year 10 (Fig. 10.4). This increase in the US soybean supply is shared between the domestic and export markets. Small initial changes in supply can readily be absorbed by the domestic market, but as supply continues to expand, a larger share of that supply is directed to the export market. At full adoption, US soybean exports would grow by 2.0%. Soybean co-product supply would be similarly affected. Soybean meal production would rise by roughly the same amount as domestic consumption of soybeans (1.75%), as virtually all soybeans are crushed for oil and meal. Soybean meal exports would then expand, as the domestic demand is relatively inelastic due to competition from other protein meals, primarily distillers' dried grains. US exports of soybean meal would therefore rise by more than 4% above baseline

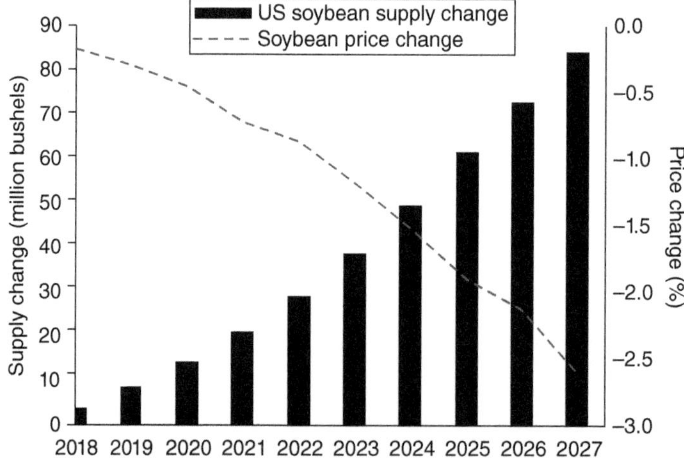

Fig. 10.4. Changes in US soybean supply and soybean prices. (Data from author models.)

at full adoption. Soybean oil production would also rise, but virtually all extra soybean oil produced would be retained by the domestic market.

Increases in soybean supply would be met with decreases in soybean price (Fig. 10.4). The decrease in soybean price is just over 2.5% below baseline when the innovation is fully adopted, in year 10. Changes in the soybean price impact many aspects of the agricultural economy, from which crops farmers choose to plant to how much is consumed. Lower prices reduce gross returns for all soybean farmers, while yield gains and cost-efficiencies increase returns for adopters.

The decline in soybean price induces the prices of competitive oil crops to fall as well. Under the innovation scenario, canola, sunflower, and palm oil prices would decline by 0.8%, 0.4%, and 0.3%, respectively (Fig. 10.5). The prices of grains, maize and wheat, for example, would also drop slightly because of small increases in their acreage, which occur as their relative profitability increases against that of soybeans in many producing countries where farmers experience falling soybean prices.

On the whole, the novel resistance trait would create efficiencies that benefit society; the distribution of these benefits is determined by the changes in market conditions outlined above. The US soybean farmers who adopt the novel technology benefit through higher yields and lower costs but transfer some of these benefits to consumers through lower soybean prices. Non-adopters become less profitable due to the decline in prevailing soybean prices, which are not offset by lower production costs or increased yields. As all international soybean farmers are non-adopters, by assumption, they are negatively impacted by the downward pressure on global soybean prices. Farmers growing other crops that experience declining prices are also negatively impacted. The declines in soybean and other crop prices, however, are realized as benefits to consumers globally.

Table 10.1 reports the estimated global benefits from adoption of the novel broad resistance trait, expressed as the total changes in producer and consumer surplus over the period 2018–2027. It also reports the estimated gross revenue of the technology suppliers. Together, these figures represent the total estimated

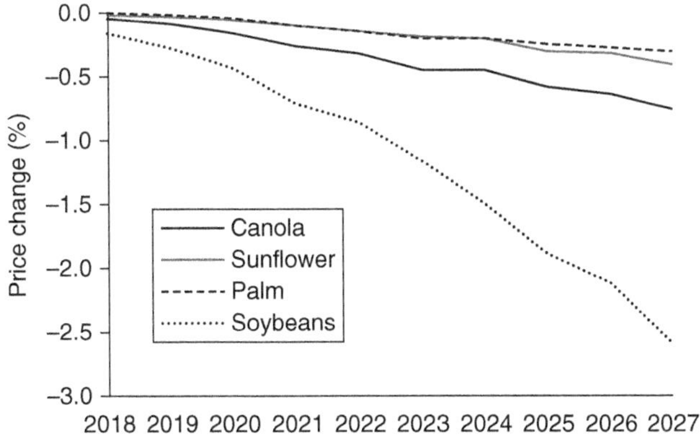

Fig. 10.5. Changes in prices of selected crops. (Data from author models.)

Table 10.1. Change in producer and consumer surplus from adoption of a broad resistance trait from 2018 to 2027 (prices in $1000). (Data from author models.)

Country/region	From changes in soybean markets		From changes in all crop markets		Total change in surplus
	Producer surplus	Consumer surplus	Producer surplus	Consumer surplus	
Canada	−538	153	−894	332	−562
Mexico	−20	235	−56	323	267
USA	223	2,927	−174	3,252	3,078
Argentina	−3,119	2,598	−3,192	2,643	−549
Brazil	−5,884	2,541	−6,039	2,625	−3,413
Paraguay	−458	178	−463	181	−283
European Union (28 members)	−128	734	−753	1,478	725
Russia	−202	293	−385	468	83
Ukraine	−232	88	−422	229	−193
Middle East	−12	296	−69	461	392
China	−751	6,113	−1,226	6,794	5,568
Japan	−12	154	−13	218	205
Taiwan	0	108	0	116	116
India	−597	592	−830	977	147
Pakistan	0	166	−29	270	241
Indonesia	−27	155	−596	335	−262
Malaysia	0	39	−348	100	−248
Thailand	−2	150	−53	191	138
Vietnam	−8	111	−10	139	128
North Africa	−1	183	−14	249	235
Sub-Saharan Africa	−133	138	−275	375	100
Rest of the world	−381	436	−621	713	92
World	**−12,283**	**18,387**	**−16,463**	**22,469**	**6,006**
Tech supplier gross return					**1,338**
Total global surplus					**7,344**

benefits from adoption of the conjectural novel trait over the period of analysis. More disaggregated figures on the changes in producer and consumer surplus for selected countries and regions are reported in Table 10.2, in order to clarify the cross-commodity impacts of the innovation.

Global consumer surplus increased by almost US$22.5 billion, comprising US$18.4 directly from the reduction in soybean prices and the rest from the net effect of the induced changes in prices and consumption of all other oilseeds and grains.[37] Large gains in consumer surplus were realized by large consuming countries such as China, the USA, Brazil, the EU, India, Japan, and Argentina. However, in the case of Argentina, while much of the soybean production is processed domestically, value-added products (e.g. soybean meal, soy oil) are exported to many countries, such that much of the consumer surplus gain accrues to consumers in importing countries.

Table 10.2. Change in producer and consumer surplus in selected crops from adoption of a broad resistance trait from 2018 to 2027 (prices in $1000). (Data from author models.)

Country/region	Soybeans		Maize		Canola		Cotton		Palm oil		Wheat		All crops	
	PS	CS	PS	CS	PS	CS	PS	CS	PS	CS	PS	CS	PS	CS
Canada	−538	153	−10	11	−329	156	0	0	0	2	−10	3	−894	332
Mexico	−20	235	−20	34	0	29	−1	1	−2	10	−1	3	−56	323
USA	223	2,927	−316	230	−29	34	−14	1	0	20	−12	10	−174	3,252
Argentina	−3,119	2,598	−30	11	−1	0	0	0	0	0	−5	2	−3,192	2,643
Brazil	−5,884	2,541	−84	53	0	0	−3	1	−6	10	−2	4	−6,039	2,625
Paraguay	−458	178	−3	1	−1	1	0	0	0	0	0	0	−463	181
European Union (28 members)	−128	734	−46	58	−372	404	−1	0	0	92	−52	41	−753	1,478
Russia	−202	293	−12	7	−21	20	0	0	0	12	−24	14	−385	468
Ukraine	−232	88	−20	5	−28	5	0	0	0	0	−9	3	−422	229
Middle East	−12	296	−7	23	−6	24	−2	3	0	36	−14	22	−69	461
China	−751	6,113	−174	192	−230	322	−10	14	0	82	−38	40	−1,226	6,794
Japan	−12	154	0	11	0	39	0	0	0	10	0	2	−13	218
Taiwan	0	108	0	3	0	0	0	0	0	4	0	0	0	116
India	−597	592	−23	22	−138	138	−13	10	−3	157	−32	33	−830	977
Pakistan	0	166	−5	5	−3	23	−4	5	0	53	−8	9	−29	270
Indonesia	−27	155	−9	11	0	0	0	1	−564	161	0	3	−596	335
Malaysia	0	39	0	3	0	0	0	0	−348	55	0	1	−348	100
Thailand	−2	150	−4	4	0	0	0	0	−35	34	0	1	−53	191
Vietnam	−8	111	−4	11	0	0	0	2	0	14	0	1	−10	139
North Africa	−1	183	−4	19	0	0	0	0	0	22	−4	12	−14	249
Sub-Saharan Africa	−133	138	−56	61	−1	1	−3	1	−43	116	−3	11	−275	375
Rest of the world	−381	436	−35	57	−69	32	−4	5	−66	113	−19	27	−621	713
World	−12,283	18,387	−862	833	−1,228	1,228	−54	47	−1,067	1,002	−235	241	−16,463	22,469

The global loss in surplus among soybean producers from the adoption of the novel resistance trait is estimated to be approximately US$12.3 billion over the period 2018–2027. The economic impact on soybean farmers is driven by the price effects in the global market. Soybean farmers in Brazil, Argentina, Canada, China, and India suffer the largest losses, proportional to their levels of production. Soybean farmers in the US benefit from the new technology by gaining almost $225 million, as a group. However, this figure obscures an important divide. US adopters of the novel trait would gain approximately $2.75 billion over the 10-year period of the analysis, while non-adopters would experience a loss of surplus of just over $2.5 billion.

Producer surplus for the farmers of other oil crops also declines because they face lower prices without any offsetting reductions in their variable costs or yield gains. Combining sunflowers, canola and palm oil, the total loss in producer surplus for these crops is US$2.7 billion. Similarly, producer surplus for other crops (e.g. maize, cotton, wheat) falls by a total of US$1.5 billion, once again because of price declines resulting from their acreage and supply increases (Table 10.2).

Technology suppliers realize a total gross revenue of approximately $1.34 billion from sales of the novel trait over the 10-year period of 2018–2027. This amount of surplus derived from the innovation is transferred from adopters to technology suppliers through the $5/acre farmer payment for the novel seed. This is a gross gain for the technology supplier as it does not account for expenses incurred in the development (e.g. R&D, regulatory) and marketing of the innovation. It also does not account for the loss of revenue associated with the payments for seed treatments, partial resistance, and *Rps* genes used in the baseline scenario that are replaced by the novel trait. Indeed, under the assumptions in our innovation scenario, the technology suppliers, as a group, experience a net loss of revenue over the period of analysis.[38]

Overall, the total global benefits from the novel resistance trait are approximately $7.3 billion over the 2018–2027 period. From these, US soybean farmers who adopt the technology receive $2.75 billion (37.5%), the technology suppliers receive $1.35 billion (18.3%), and the remaining $3.2 billion (43.5%) accrues to consumers across the world distributed through trade. Consumers also gain another $19 billion, which represents a net transfer from farmers of soybeans and other crops to consumers around the world.

The temporal pattern of innovation also affects the size and distribution of the associated benefits. As adoption grows over time, the price changes induced by the innovation across commodity markets, as well as the changes in consumer and producer surplus, vary from year to year. Figure 10.6 illustrates the year-to-year variations in the changes of US producer and consumer surplus, as well as in the fees paid to the technology provider. As adoption grows, the benefits of innovation grow in size, and a progressively larger share of the benefits is directed to consumers. Indeed, although the period of analysis ends in 2027, some $1 billion/year in welfare gains from the innovation are realized for as long as the novel trait continues to be used by US soybean farmers.

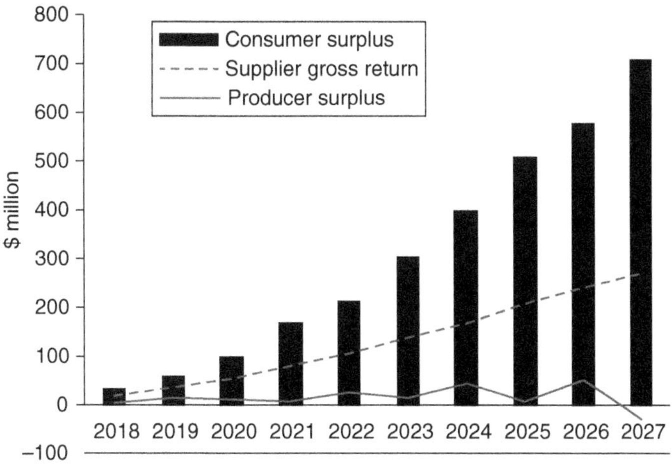

Fig. 10.6. Changes in benefits to consumers, producers, and technology suppliers. (Data from author models.)

Summary

Our analysis in this chapter reinforces certain results and lessons learned from previous studies in agricultural innovation. First, adopters profit from the innovation while non-adopters and farmers of competitive crops who do not benefit from parallel cost-efficiencies and yield gains experience economic losses. Second, through induced reductions in market prices, consumers benefit the most from the innovation. Because soybeans and other agricultural commodities are widely traded, a large share of the benefits is transferred to consumers in importing countries. Third, the aggregate economic impacts of the novel trait are large and sustained but also quite dynamic, and are shaped by the patterns of adoption and conditions in global commodity markets.

Our analysis also provides some new insights and clarifies certain results in Chapters 8 and 9 (this volume). As we discussed in Chapter 8, most soybean farmers with disease on their farm would profit from and adopt the novel broad resistance trait, even if it was priced high. Nevertheless, because the technology suppliers cannot discriminate among soybean farmers with high and low disease incidence, they must offer the novel trait at a low enough price to encourage broad adoption. This difference in the implied economic value of the novel trait for certain farmers and its market price is the source of the producer surplus enjoyed by the adopters. This market mechanism limits the potential profitability of the innovation to the technology supplier but expands the economic benefits to the adopters. The temporal pattern of the revenue received by the technology supplier also reveals that the bulk of such income comes well into the adoption path, when the new trait is employed on a larger number of acres. Hence, accounting for the more than 10 years of R&D that characterize the development of such innovations as the conjectural broad resistance trait, technology suppliers must expect long lags in receiving significant pay-offs from their innovations. These market effects, along with the limited and fragmented nature of disease control input markets (see Chapter 9), can explain the observed paucity of R&D investment and product pipelines in disease control inputs.

Notes

[35] The measure of gross annual benefits in Eqn 10.5 is similar to the concept of gross annual research benefits introduced by Griliches (1958), defined as the change in the total value of production attributable to agricultural research. This measure was formalized by Alston *et al.* (1995) as the annual gross value of production multiplied by the portion of the annual change in total factor productivity attributed to research activities.

[36] Based on assumptions about the efficiency of the broad resistance trait and the current losses to seedling disease, we calculate avoided losses to be 90% × 24% × 8.2 bushels = 1.77 bushels/acre. Based on assumptions about the efficiency of the broad resistance trait and the current losses to mid-season root rots, we calculate avoided yield losses to be 90% × 22.7% × 7.8 bushels = 1.59 bushels/acre.

[37] We restrict our analysis of the welfare effects of the innovation to considering the market effects in all major crop markets. It is possible to consider the market effects in downstream markets where such crops are sold (feed, livestock, food), but such analysis is beyond the scope of this chapter. Irrespective, the total benefits would not change in size, but the distribution of these benefits would.

[38] It is important to note that such presumed losses in revenue assume that fungicidal seed treatments, partial resistance, and *Rps* genes would continue to control seedling disease and mid-season root rots during the 10-year period of analysis. If resistance build-up rendered any of these technologies ineffective, sales of such disease control inputs would not occur and revenues from the novel trait would represent a net gain.

11 Summary and Conclusions

Soybeans are both one of the oldest and one of the newest crops in the human agricultural inventory. They have been grown for thousands of years, yet only since the 1970s have soybeans become a major commodity produced in many different countries and widely traded on global markets. Soybeans are now the world's premier oilseed crop, with a multitude of food, feed, and industrial uses. Demand for soy-based products is only expected to grow in the future, presenting soybean farmers with the challenge of continually improving the productivity of land, labor, and other inputs in order to keep up.

One way to increase soybean production, given a constant resource base, is to decrease losses from pests, including pathogens. With estimates of global production losses from disease in the 10–20% range occurring every year, there is scope to increase soybean yields through better disease control. To improve disease control, soybean farmers must adopt more advanced practices and use them effectively in their fields. In turn, input suppliers must develop improved disease control practices and inputs and enable farmers to adopt them. The purpose of this book was to examine the economic decisions of soybean farmers and input suppliers in disease management and the factors that shape them. Understanding these economic decisions is essential as they – ultimately – determine whether disease control in soybean production can improve and whether the productivity challenge can be met.

Summary of the Main Points

Throughout the book, we have used concepts and methods that are broadly applicable to the management of all soybean diseases and, indeed, all crop diseases. When we needed to examine the disease management process in detail, we focused our analysis on control of a specific pathogen class (oomycetes) in a given geographical area (the USA).

Combatting pathogens presents a complex economic problem for soybean farmers due to the nature and level of uncertainty involved. Disease diagnosis can be challenging; infection does not always produce visible symptoms, and symptoms do not always definitively indicate the causal pathogen. More broadly, anticipating the potential incidence and severity of disease and the need for management is inherently difficult. Whether any specific disease develops in any given field and year depends on the presence of a virulent pathogen, a susceptible host, and environmental conditions favorable to infection. The pathogen–host–environment interplay, however, is complex, and the outcomes are difficult to predict. For instance, the population size and profile of oomycetes, as well as other pathogens, in any particular field can vary substantially from year to year, and those in different fields can vary independently. Agronomic practices can have a strong impact on pathogen populations. Improving soil drainage and optimizing irrigation can limit the saturated soil conditions necessary for oomycete reproduction. Tillage breaks up hyphal masses and can bury spores below the root zone. Crop rotation can rob oomycetes of susceptible host plants for a season, and shifting planting dates can introduce the host crop when the pathogen is less virulent. Changing environmental conditions can also dramatically alter the path of disease progression in any given year. Annual weather changes help determine soil moisture and temperature at planting time, and thus both oomycete virulence and soybean susceptibility. This influence is mediated by the type of soil in a particular field, as water capacity and warming speed vary with soil types. Against this backdrop of inherent variability and unpredictability, farmers must make disease treatment decisions based on expectations of incidence and severity, often formed with sparse information.

The farmer's disease control decisions can be described by a damage abatement model where the farmer maximizes profits (or utility) by balancing the cost of each disease control practice against the expected value of the production loss it prevents. The value of production loss prevented by a particular control practice that is equal to the cost of that practice is defined as its economic threshold (ET). The farmer will take action and employ the practice only if the expected production loss meets or exceeds the ET. Otherwise, it is more economical to simply suffer the loss.

In order to estimate ETs and formulate a disease control plan through damage abatement considerations, farmers need reasonably accurate expectations about the incidence and severity of disease in their individual fields. There is, however, no extant data on such expectations and no simple way of measuring them, as they are largely subjective. We expect a fairly close correlation between farmers' expectations and actual historic incidence and severity on the farm, but we also know that various factors can distort perceptions. Thus, as a first step for our study, we used surveys to elicit farmers' experiences and expectations of seedling disease and mid-season root rot occurrences in their fields. In the absence of any baseline data on actual incidence and severity, we also surveyed expert CCAs to discover their experience with disease in their areas of operation. Our results indicated that farmers' expectations of disease incidence, severity, and loss in soybean production are comparable to those of experts and consistent with reported aggregate production losses. We therefore concluded that, despite the demanding nature of the information they need to collect, soybean farmers form effective expectations of disease incidence and severity in their fields.

When farmers expect that disease incidence and severity will be at a level that requires action, they have three types of control practices available: agronomic, chemical, and genetic. Each practice comes with its own set of costs and benefits, but there are important tradeoffs that must be accounted for. In the context of seedling disease and mid-season root rots, for example, conventional tillage can break up fungal and oomycete masses and bury them below planting depth, decreasing disease prospects. Reduced tillage, in contrast, offers cost savings in terms of fuel and machinery time, as well as environmental benefits such as decreased runoff and soil erosion, but might increase disease incidence and severity. Similarly, earlier planting increases yield by extending the growing season, but planting in cooler, wetter soil also increases the probability of disease. Chemical fungicides are available as seed treatments, foliar sprays, and, increasingly rarely, soil treatments with different efficacies and costs. Genetic resistance comes in two varieties. Specific resistance grants near-total immunity but only to pathogen races that contain an avirulence gene that matches the resistance gene in the host. Partial resistance operates against an entire pathogen species and substantially reduces the severity of disease but does not offer complete protection. A specific resistance trait may or may not be available in seed varieties that are preferred for yield, maturity date, or other characteristics. Farmers must therefore consider the relative efficacy, cost, and benefit of each available control practice, as well as the relevant tradeoffs, in order to choose the practices that match their specific agronomic and disease management needs.

The relative effectiveness of a disease control practice is a key consideration; formulating reasonable expectations of effectiveness is essential to a disease management program. To understand the accuracy of farmers' expectations, we used surveys to elicit their perceptions of the relative effectiveness of practices used to control seedling disease and mid-season rots. We also asked them to report which of these practices they actually used on their farms. In addition, we surveyed expert CCAs to create baseline data on the actual effectiveness of alternative practices used for seedling disease and mid-season rot control. Our survey results indicated that farmers' expectations of the effectiveness of various control practices are similar to those of experts. We therefore concluded that, despite the demanding nature of the information they need to collect, soybean farmers form accurate expectations about the relative effectiveness of alternative disease control practices they might use.

Our survey results also indicated that farmers act on their expectations. In particular, we found that some agronomic practices, specifically early planting and conservation tillage, offer perceived yield and efficiency benefits to farmers that are much greater than the disease costs they might impose. As such, these agronomic practices do not feature in farmers' considerations of disease control. Instead, they prefer chemical seed treatments and genetic resistance traits as their primary control practices. These are the most flexible practices, as they can be easily applied to any quantity of seed or number of acres.

After forming expectations regarding disease incidence and severity, the effectiveness of the various control measures, and market prices, soybean farmers choose the chemical seed treatments, genetic resistance traits, or other disease control practices that maximize their farm profits (or utility). Changes in these expected values can lead to different optimal choices. As a result, soybean farmers must frequently gather information and update their expectations in order to make effective economic decisions about disease control.

In the process of maximizing profits or utility, farmers thus form individual demand schedules for disease control measures. In fact, we found that the disease control practices that soybean farmers actually use in their fields are consistent with those that maximize profits in our stylized economic models. We therefore concluded that soybean farmers make effective economic decisions about disease control in spite of inherent uncertainty, incomplete information, and intense computational demands in their profit-maximizing calculus.

Farmers and farms are, however, heterogeneous in terms of environmental conditions, pathogen populations, farming systems, risk attitudes, and other factors. As a result, optimum choices in disease control can vary widely across farmers. These differences in the optimum choices among farmers underpin the movements in the aggregate demand for disease control inputs that occur when input prices change. In sum, each farmer's rational economic decision making, with the goal of profit maximization, is the foundation of individual and aggregate market demand for disease control practices and inputs.

To improve disease control, soybean farmers must use more efficient practices and inputs. After forming expectations about relative disease treatment efficacy and costs, soybean farmers can evaluate any innovation in disease control using their usual profit-maximization calculus. They compare the expected pay-off from the innovation with the pay-offs from alternative practices and inputs; when the innovation is the option that maximizes profits, they adopt it. The profit-maximizing pay-off of the innovation defines its intrinsic economic value. Soybean farmers are willing to pay a price for the innovation that is as high as its per-unit pay-off but no more. Because of the inherent heterogeneity, the economic value of the innovation varies across farms and farmers. As a result, different farmers have different WTP for any innovation in disease control. The distribution of the WTP values of all farmers defines the market demand for the innovation.

When we presented the possibility of a novel broad genetic resistance trait against oomycetes, soybean farmers demonstrated WTP for the novel trait that were comparable to those expected by experts and consistent with our stylized economic models. We therefore concluded that soybean farmers make effective economic decisions when they consider the potential adoption of new practices and inputs for disease control. Depending on the circumstances on their farm but also on their personal preferences, soybean farmers demonstrated a wide spread in their WTP for the new genetic resistance trait. In this context, we illustrated that the level of adoption for such an innovation would depend ultimately on farmer demand but also on the price at which it would be offered by suppliers.

Whatever the farm demand for chemical controls, genetic resistance traits, or other disease control inputs might be, these inputs must first be developed and commercialized by the agricultural input industry. R&D efforts in this industry have been helped by constant technical progress, but more stringent regulations have slowed product development and increased costs. Pathogen adaptation and resistance build-up to commonly used fungicides or genes has also forced input suppliers to develop replacement products with new and multiple modes of action. The most significant challenge faced by input suppliers in the development of innovations for disease control, however, seems to be securing sufficient pay-offs to justify the required R&D investment. Unpredictable and intermittent incidence and severity of disease tend to limit FWTP for disease control. What demand exists is fragmented, as pathogens are different and require separate remedies. Often then, there is a wedge between the prices farmers are willing to pay for disease control inputs and the prices suppliers believe they must charge in order to recoup R&D costs and fund future innovation. This explains our finding that the existing disease control input inventory, both chemical and genetic, is small and dated, and that the R&D pipeline is sparsely populated.

The lack of strong innovation is unfortunate, as higher yields advance the sustainability of soybean production, and the socio-economic benefits from improved disease control are large and broadly distributed among producers and consumers. Reduced yield losses to disease and increased profit per acre translate directly into greater supply in global soybean markets and correspondingly lower soybean prices. The lower prices distribute the benefit to the buyers of soy products and ultimately to consumers around the world. Consumers unambiguously benefit from lower prices, and the farmers who adopt the innovations gain when the value of the yield and cost-efficiencies exceed the loss of revenue from the lower prices. Economic gains to society from innovation in soybean disease control can readily be quantified, as in the case of the broad resistance trait against seedling disease and mid-season root rots that we studied in this book. Assuming only gradual adoption over time, diffusion of this trait to US farmers alone could produce economic benefits to society in excess of $7 billion over a 10-year period, most of which accrue to consumers.

Future Developments

The brief summary above highlights the basic economics of soybean disease management in its present state. There are, however, important developments that may bring about significant change in the near and distant future that deserve attention. These include climate change and a flow of fundamental technical innovations, as well as significant restructuring in the agricultural input industry. We briefly discuss each in turn.

Climate change and soybean diseases

Given the strong influence that the environment and weather patterns have on plant diseases, the prospect of global climate change holds implications for disease management in soybean production that cannot be ignored. Climate change can potentially impact the incidence and severity of diseases through a few different avenues, all of which may have variable and hard-to-predict effects on production losses.

Climate change is expected to bring about an increase in global average temperature, identified previously as one of the most important environmental variables affecting diseases. Warmer weather in general is likely to favor some pathogens and inhibit others, depending on what coincident changes might happen with humidity and rainfall. Warmer and moister conditions could increase problems from most fungal infections, while warmer and drier times may inhibit these and foster others, such as powdery mildew, that prefer less moisture (Rosenzweig *et al.*, 2001). Additionally, warmer temperatures will shift all climate zones toward the poles, and with them the ideal environments for particular soybean varieties, disease pathogens, and invertebrate vectors (Anderson *et al.*, 2004). These various environments may not shift in concert, which could change the incidence and severity of some diseases in specific locales. In addition, as the zones move, they may expand into different physical environments in terms of soil characteristics, terrain, and native vegetation, all of which could effect changes in the growth and physiology of host crops and pathogens alike. The resulting disease conditions are likely to vary considerably across locations and be very difficult to predict (Chakraborty *et al.*, 2000).

The purported driver of climate change and its attendant temperature effects is an increase in atmospheric CO_2 and other greenhouse gases. Higher CO_2 concentrations can have direct effects on plant growth, in addition to the indirect changes from higher temperatures. These changes likewise place conflicting pressures on disease incidence and severity. On the one hand, increased CO_2 levels affect leaf physiology in ways that can increase disease resistance, including changes to stomata and thicker wax layers and cuticles. On the other hand, more CO_2 also promotes earlier and more vigorous growth. Earlier growth can make young plants more attractive targets for infection, and a denser canopy makes the underlying microenvironment warmer and more humid, leaving the plants more vulnerable to soil pathogens that attack the stem and roots (Chakraborty *et al.*, 2000). The end results on disease are again variable and hard to predict. Increased CO_2 levels have been connected with a decrease in severity of downy mildew of 40% or more, a slight increase in brown spot severity, and no change in SDS. In studies of the impact of doubled CO_2 concentration on 25 pathogens, 15 exhibited increased severity, eight showed reduced severity, and two did not change (Luck *et al.*, 2011).

Most climate models also predict an increase in extreme weather events such as droughts, floods, and storms as climate change progresses. The main outcome of these events will likely be an increase in the variability of disease conditions as the weather changes to favor one or another pathogen. In addition, drought and other extreme weather can stress plants and make them more susceptible to infection, while floods and storms can facilitate the spread of disease into new territories, as described earlier (Rosenzweig *et al.*, 2001). Hurricanes and the continental weather patterns they spawn have already been documented to have spread pathogens into new territories, including bringing soybean rust to North America (Luck *et al.*, 2011).

Finally, climate change can spur the emergence of new diseases, regardless of exactly what the new climatic conditions turn out to be. As climate zones shift geographically, not only will crops and pathogens adapt to the new environment, as noted above, but different pathogens will come into contact with one another that otherwise would not have, creating new opportunities for pathogen hybridization (Anderson *et al.*, 2004).

Significant changes in disease incidence and severity and the potential introduction of new pathogens would require farmers and input developers alike to adapt over relatively short timespans. A nimble, resourceful, and responsive R&D system would be best situated to deal with unexpected new pathogens and new conditions. As we have seen, however, the current innovation system has none of these characteristics. The R&D process takes many years and millions of dollars to yield marketable new products, and adaptation of existing products to new geographies and growing conditions can be just as laborious. Nevertheless, there are important technical innovations under way that, over time, could change the current fundamentals of soybean disease management.

Emerging technical innovations

In addition to the fundamental discoveries in gene editing discussed in Chapter 9 (this volume), there is a suite of technical innovations currently under way that is promising sweeping changes in agricultural production and disease control.

Digital agriculture

The emergence of digital agriculture (DA) has been many years in the making. DA combines long-known precision agriculture technologies such as yield monitors, variable rate technologies, and GPS guidance and control systems with newer ones like remote sensing, robotics, autonomous vehicles, cloud-based data sharing and computing, artificial intelligence, machine learning, and a host of decision support tools. The result is a system that is broadly applicable to all types of agricultural production, including soybean production (van Es *et al.*, 2016). DA capabilities hold promise for reducing uncertainty and improving control in disease management.

As discussed throughout this book, soybean diseases are inherently difficult to predict and manage due to the complexity of the disease process. DA data tools are well suited to address this problem. There have already been a number of approaches to monitor and predict disease incidence both at the field level and at a larger regional level (Donatelli *et al.*, 2017). Integrated models can overlay various data points with relevance to diseases, including rainfall, soil moisture, temperature, soil type, soil drainage, crop type, specific disease history, and various others (van Evert *et al.*, 2017). These models can then predict disease incidence, severity, and production losses at the field and subfield level, and can assist in disease management. More recently, an ever-increasing flow of remote sensing data has improved the capabilities of predictive models (Mulla, 2013). Remote sensing data from satellites, manned aircraft, drones, and various ground-based sensors have supplied information on land topology and soil qualities, vegetative cover, plant stressors, and other variables relevant to the analysis of disease incidence and severity (Mahlein, 2015).

Indeed, as the number and types of farm sensors have continued to expand at an exponential rate, the amount and frequency of available remote sensing data have burgeoned. Predictive analytics are greatly enhanced when computation moves to "the cloud," combining data from many fields, farms, and regions, and over time. Such big data aggregations allow structured as well as artificial intelligence and machine-learning-based models to tease out more subtle relationships among agronomic practices, farming environments, and stressors with agricultural outcomes such as disease damage (Erwin, 2016). As such, DA can assist with optimal choice and placement of seed technologies to prevent disease in fields or parts of fields but also with more timely, targeted, precise, and economical responsive treatments, such as precision sprays using drones.

DA can also enhance the effectiveness of regional and national early warning networks on disease incidence. The National Plant Diagnostic Network identified the arrival of Asian soybean rust in the USA and was able to begin the process of alerting stakeholders and monitoring the spread of the disease (Magarey *et al.*, 2009). A more recent example is the USAblight monitoring network for late blight in tomatoes and potatoes (http://usablight.org), which illustrates the ways human and machine networks and big data can combine to improve early forecasts of potential disease spread (Fry, 2016). While DA is still in the early stages of development, ongoing innovations promise to give farmers the ability to make more informed disease management decisions by reducing uncertainty and to take targeted action against diseases in new, more efficient ways.

New diagnostic tools

When diseases are present in the field, they are not always easy to diagnose, and without a proper diagnosis, effective treatment is unlikely. Disease diagnosis depends on tissue and soil sampling and laboratory testing. As a result, it is currently slow and relatively expensive. A variety of new diagnostic tools are making this process easier and faster. A marriage of immunology and nanotechnology may soon provide farmers with simple field diagnosis devices that give results in real time (Kashyap *et al.*, 2016). Machine learning and artificial intelligence embedded in smartphone apps are also being developed to diagnose crop disease in remote areas from photographs and other data (Elliot, 2016; Gill, 2016). These and other similar advances may enable farmers to make timelier and better-informed treatment decisions. Our survey results indicate that such innovations could be widely adopted, depending on price.

Biopesticides

Biopesticides are crop protection products based on microorganisms, biochemicals produced from biological sources, microbials, and products made from other natural sources (Barratt *et al.*, 2018). A major attribute that differentiates biopesticides from synthetic pesticides is their mode of action. For instance, while most, if not all, synthetic insecticides are neurotoxic to pests, many bioinsecticides have other modes of action including mating disruption, anti-feeding mechanisms, suffocation, and desiccation (Copping and Menn, 2000).

There is a long-standing tradition of research into biopesticides for crop protection, and in recent years, interest has increased because of the lower regulatory burden for bringing new products to the market (Lacey *et al.*, 2015) and the potential for development of new modes of action. Technical advances such as the decreasing cost of genetic sequencing of plant and soil microbes, more cost-effective industrial-scale fermentation processes, and the emergence of novel gene-editing and RNAi technologies have also contributed to the resurgent interest (Calvo *et al.*, 2014). Particular research attention has been paid to the development of biopesticides that can be used in combination with synthetic crop protection products and, especially, in seed treatments (Kalaitzandonakes and Zahringer, 2018). A significant share of the biochemicals used today for crop protection come from bacteria, but R&D in new product concepts has a broader base of potential sources.

Structural Changes in the Input Sector

Fundamental changes in agricultural technologies have often been paralleled by significant structural changes in the agricultural input sector. During the early days of biotechnology development in the 1980s and 1990s, a large number of mergers and acquisitions (M&As) sought to bring together the development of biotech traits and elite seed germplasm in order to accelerate commercialization (Kalaitzandonakes and Bjornson, 1997). This restructuring of the agricultural input sector produced a handful of large, vertically integrated, multi-national R&D firms with significant presence in the crop protection, biotech, and seed industries. In the last 2 years, high-profile M&As among these large multi-national R&D firms have further restructured the agricultural input sector. The merged input suppliers seem to have adopted an R&D and business model that brings together multiple R&D platforms (biologicals, synthetics, germplasm, biotech traits, and DA) in order to produce technology bundles that can maximize yields and cost-efficiencies in crop production (Kalaitzandonakes and Zahringer, 2018). For instance, seed treatments may combine multiple synthetic chemicals and biologicals in order to protect crops from insects, pathogens, and other pests while enhancing fertility and nutrient availability. These seed treatments may then be paired with superior genetic traits that have been developed for native resistance to other pathogens or modified with biotech traits to assist with limited moisture, insect resistance, and weed control through herbicide tolerance. DA and precision farming can, in turn, ensure the compatibility of planted seeds with the soil and the overall environment, and can inform the farmer in making the optimum seed variety choice. In effect, this expansive R&D and business model calls for the integrated use of multiple vertical technology platforms in the development of technology bundles with maximum yield and cost-efficiency potential.

The recent high-profile M&As are only the most visible part of industry efforts to implement this integrated technology R&D model. Firms across the agricultural input industry have implemented large numbers of licensing and marketing agreements, as well as strategic research alliances, in order to accelerate innovation through these multiple R&D platforms (Kalaitzandonakes *et al.*, 2010, 2018). Furthermore, institutional investors have funded record numbers of new startup firms that specialize in R&D in these advanced technology platforms (AgFunder, 2017). Of course, only time will tell if the full promise of these efforts will be realized.

Conclusion

While our analysis has focused on soybean disease management, it is likely that much of what we have described in this book also applies to other crops. This might be especially so for smaller-acreage crops. Indeed, a general conclusion that can be drawn from our work is that public investments and institutional reforms that advance innovation and improve farm efficiency in disease control for soybeans and other crops might be justified on economic grounds.

Increased public funding for basic research in such areas as plant pathology, the impact of climate change on pathogen–crop interactions, and other areas could encourage follow-on applied R&D in the private sector to replace the ageing disease control technologies used today and expand the portfolio of solutions. Given the broad socio-economic benefits that could accrue to consumers from more effective disease control in soybean production, private R&D efforts could lead to innovations with high social rates of return to the underlying basic research investments.

Encouraging private-sector R&D to produce disease control practices and inputs should be a priority, as the current pipeline may not even prove sufficient to replace commercial products that could be lost to pest resistance. High R&D costs as well as limited and uncertain returns to new inputs for disease control are creating a situation similar to that of "orphan" crops, where larger direct public investments have been advocated to jump-start the private R&D process and ultimately generate positive returns (e.g. Takeshima, 2010).

Institutional reform may be more difficult but could have a more significant overall impact. Regulatory oversight of modern technologies in agriculture has ensured their safety for many decades. However, the regulatory approval process for crop protection products, whether synthetic or genetic in nature, has become expensive and slow (Kalaitzandonakes *et al.*, 2007). This increase in fixed compliance costs places an additional hurdle in the path of innovations on the way to commercialization. For products in smaller, fragmented markets, as most crop disease control markets are, the barrier is often insurmountable. As such, regulatory costs may prevent the development of innovations that could be profitable under a less restrictive regime. Thus, regulatory restrictions also create a situation akin to that of orphan crops, described above. Various reforms have been proposed, including adapting the extent of required testing to the anticipated risks and using risk modeling to decrease data needs (Falck-Zepeda and Cohen, 2006), or even fundamentally changing the structure of regulations away from the current event- or molecule-specific regime to something more generalized (Bradford *et al.*, 2006). Regulatory uncertainty can have nearly as chilling an effect on innovation as known high costs. Firms may be slow to invest in certain R&D projects if the regulatory status of a class of innovations is likely to change in the near future. This is a major consideration for gene-editing technologies, such as CRISPR/Cas9, as the nature of future regulatory oversight remains uncertain (Hartung and Schiemann, 2014; Jones, 2015).

Beyond new control practices and inputs, access to high-quality and timely information on disease incidence and severity and ways to control them can help farmers improve their decision making and level of effectiveness in disease management. Farmers in the USA have access to multiple sources of high-quality information, and our survey results indicated that they use these information sources effectively. However, for many farmers around the world, such access is not currently possible. This may be another area where public investments could have results with large and widespread social benefits. Public investments in modern communications technologies (Aker, 2011), broader data generation (Falck-Zepeda and Cohen, 2006), and other similar activities could enable better decision making on the part of farmers in both developed and developing countries.

There has been little attention paid to the economics of crop disease management in the past. The relevant academic research is scant and so is the public understanding of the dated practices for controlling disease, the growing pathogen resistance to the few available controls, the feeble innovation pipelines, and the inherent uncertainty of potential climate change. Our hope, then, is that our work in this book will motivate additional research and badly needed attention to these important challenges that could limit crop and food supplies in the future, just when the world needs them most.

References

Abdelmajid, K.M., Meksem, K., Wood, A.J., and Lightfoot, D.A. (2007) Loci underlying SDS and SCN resistance mapped in the 'Essex' by 'Forrest' soybean recombinant inbred lines. *Reviews in Biology and Biotechnology* 6, 2–10. https://doi.org/10.1023/B:PLSO.0000030189.96115.21

Affholder, F., Poeydebat, C., Corbeels, M., Scopel, E., and Tittonell, P. (2013) The yield gap of major food crops in family agriculture in the tropics: assessment and analysis through field surveys and modelling. *Field Crops Research* 143, 106–118. https://doi.org/10.1016/j.fcr.2012.10.021

AgFunder (2017) AgTech investing report: year in review 2016. Available at: https://agfunder.com/research/agtech-investing-report-2016 (accessed 20 September 2018).

Aggarwal, P.K., Kalra, N., Chander, S., and Pathak, H. (2006) InfoCrop: a dynamic simulation model for the assessment of crop yields, losses due to pests, and environmental impact of agro-ecosystems in tropical environments. I. Model description. *Agricultural Systems* 89, 1–25. https://doi.org/10.1016/j.agsy.2005.08.001

Agrios, G.N. (2005) *Plant Pathology*, 5th edn. Elsevier Academic Press, London.

Aker, J.C. (2011) Dial "A" for agriculture: a review of information and communication technologies for agricultural extension in developing countries. *Agricultural Economics* 42, 631–647. https://doi.org/10.1111/j.1574-0862.2011.00545.x

Alexandratos, N. and Bruinsma, J. (2012) *World Agriculture Towards 2030/2050: the 2012 Revision*. ESA Working Paper No. 12-03. FAO, Rome, Italy.

Alston, J.M., Norton, G.W., and Pardey, P.G. (1995) *Science Under Scarcity: Principles and Practice for Agricultural Research Evaluation and Priority Setting*. Cornell University Press, Ithaca, New York.

Alston, J.M., Kalaitzandonakes, N., and Kruse, J. (2014) The size and distribution of the benefits from the adoption of biotech soybean varieties. In: Smyth, S.J., Phillips, P.W.B., and Castle, D. (eds) *Handbook on Agriculture, Biotechnology, and Development*. Edward Elgar, Cheltenham, UK, pp. 728–751.

Ames, K.A., Joos, D.K., and Nafziger, E.D. (2016) *Soybean Variety Test Results in Illinois – 2016*. Crop Sciences Special Report 2016-04. University of Illinois, Urbana, Illinois. Available at: http://vt.cropsci.illinois.edu/soybean16/123499%202016%20Soybean%20booklet.pdf (accessed 20 September 2018).

Amtmann, A., Troufflard, S., and Armengaud, P. (2008) The effect of potassium nutrition on pest and disease resistance in plants. *Physiologia Plantarum* 133, 682–691. https://doi.org/10.1111/j.1399-3054.2008.01075.x

Anderson, J.R., Dillon, J.L., and Hardaker, J.B. (1985) Farmers and risk. Paper presented at the *1985 International Conference of Agricultural Economists*, 26 August–4 September, Malaga, Spain.

Anderson, P.K., Cunningham, A.A., Patel, N.G., Morales, F.J., Epstein, P.R., and Daszak, P. (2004) Emerging infectious diseases of plants: pathogen pollution, climate change and agrotechnology drivers. *Trends in Ecology and Evolution,* 19, 535–544. https://doi.org/10.1016/j.tree.2004.07.021

Arbuckle, J.G., Robertson, A., Kalaitzandonakes, N., and Kaufman, J. (2012) *Assessment of Certified Crop Adviser Experience with and Knowledge of Early and Mid-season Soybean Diseases Caused by Oomycetes.* Sociology Technical Reports, Paper No. 2. Iowa State University, Ames, Iowa. Available at: http://lib.dr.iastate.edu/soc_las_reports/2 (accessed 10 December 2018).

Arbuckle, J.G., Kalaitzandonakes, N., Kaufman, J., and Robertson, A. (2015) *Assessment of Certified Crop Adviser Experience with and Knowledge of Early and Mid-season Soybean Diseases Caused by Oomycetes: 2015.* Sociology Technical Report. Iowa State University, Ames, Iowa.

Arnold, R.W. (2005) Soil survey and soil classification. In: Grunwald, S. (ed.) *Environmental Soil-Landscape Modeling: Geographic Information Technologies and Pedometrics.* CRC Press, Boca Raton, Florida, pp. 37–60.

Athow, K.L., and Laviolette, F.A. (1982) *Rps6*, a major gene for resistance to *Phytophthora megasperma* f. sp. *glycinea* in soybean. *Phytopathology* 72, 1564–1567. https://doi.org/10.1094/Phyto-72-1564

Athow, K.L., Laviolette, F.A., Mueller, E.H., and Wilcox, J.R. (1980) A new major gene for resistance to *Phytophthora megasperma* var. *sojae* in soybean. *Phytopathology* 70, 977–980. https://doi.org/10.1094/Phyto-70-977

Auld, B.A. and Tisdell, C.A. (1987) Economic thresholds and response to uncertainty in weed control. *Agricultural Systems* 25, 219–227. https://doi.org/10.1016/0308-521X(87)90021-7

Barkley, A.P. and Porter, L.L. (1996) The determinants of wheat variety selection in Kansas, 1974 to 1993. *American Journal of Agricultural Economics* 78, 202–211. https://doi.org/10.2307/1243791

Barratt, B.I.P., Moran, V.C., Bigler, F., and van Lenteren, J.C. (2018) The status of biological control and recommendations for improving uptake for the future. *BioControl* 63, 155–167. https://doi.org/10.1007/s10526-017-9831-y

Bayer Crop Science (2016) Allegiance. Available at: www.cropscience.bayer.us/products/seedgrowth/allegiance (accessed 15 June 2016).

Beach, E.D. and Carlson, G.A. (1993) A hedonic analysis of herbicides: do user safety and water quality matter? *American Journal of Agricultural Economics* 75, 612–623. https://doi.org/10.2307/1243568

Benfenati, E. (Ed.) (2007) *Quantitative Structure–Activity Relationships (QSAR) for Pesticide Regulatory Purposes.* Elsevier, Amsterdam, The Netherlands.

Bernard, R., Smith, P., Kaufmann, M., and Schmitthenner, A. (1957) Inheritance of resistance to Phytophthora root and stem rot in the soybean. *Agronomy Journal* 49, 391. https://doi.org/10.2134/agronj1957.00021962004900070016x

Bernard, R.L. and Cremeens, C.R. (1981) An allele at the *rps1* locus from the variety 'Kingwa'. *Soybean Genetics Newsletter* 8, 40–42.

Boyd, L.A., Ridout, C., O'Sullivan, D.M., Leach, J.E., and Leung, H. (2013) Plant–pathogen interactions: disease resistance in modern agriculture. *Trends in Genetics* 29, 233–240. https://doi.org/10.1016/j.tig.2012.10.011

Bradford, K.J., Alston, J.M., and Kalaitzandonakes, N. (2006) Regulation of biotechnology for specialty crops. In: Just, R.E., Alston, J.M., and Zilberman, D. (eds) *Regulating Agricultural Biotechnology: Economics and Policy*. Springer, Boston, Massachusetts, pp. 683–697.

Bradley, C. (2008) Effect of fungicide seed treatments on stand establishment, seedling disease, and yield of soybean in North Dakota. *Plant Disease* 92, 120–125. https://doi.org/10.1094/PDIS-92-1-0120

Bradley, C., Allen, T., and Esker, P. (2016) Estimates of soybean yield reductions caused by diseases in the United States. Available at: http://extension.cropsci.illinois.edu/fieldcrops/diseases/yield_reductions.php (accessed 1 August 2016).

Broders, K., Lipps, P., Paul, P., and Dorrance, A. (2007a) Characterization of *Pythium* spp. associated with corn and soybean seed and seedling disease in Ohio. *Plant Disease* 91, 727–735. https://doi.org/10.1094/PDIS-91-6-0727

Broders, K., Lipps, P., Paul, P., and Dorrance, A. (2007b) Evaluation of *Fusarium graminearum* associated with corn and soybean seed and seedling disease in Ohio. *Plant Disease* 91, 1155–1160. https://doi.org/10.1094/PDIS-91-9-1155

Buzzell, R.I. and Anderson, T.R. (1981) Another major gene for resistance to *Phytophthora megasperma* var. *sojae* in soybean. *Soybean Genetics Newsletter* 18, 30–33.

Buzzell, R.I. and Anderson, T.R. (1992) Inheritance and race reaction of a new soybean *Rps1* allele. *Plant Disease* 76, 600–601. https://doi.org/10.1094/PD-76-0600

Calvo, P., Nelson, L., and Kloepper, J.W. (2014) Agricultural uses of plant biostimulants. *Plant and Soil* 383, 3–41. https://doi.org/10.1007/s11104-014-2131-8

Carrasco-Tauber, C., and Moffitt, L.J. (1992) Damage control econometrics: functional specification and pesticide productivity. *American Journal of Agricultural Economics* 74, 158–162. https://doi.org/10.2307/1242999

Carson, M.L. (1997) Plant breeding. In: Rechcigl, N.A. and Rechcigl, J.E. (eds) *Environmentally Safe Approaches to Crop Disease Control*. CRC Press, Boca Raton, Florida, pp. 201–220.

Chakraborty, S., Tiedemann, A.V., and Teng, P.S. (2000) Climate change: potential impact on plant diseases. *Environmental Pollution* 108, 317–326. https://doi.org/10.1016/S0269-7491(99)00210-9

Chamnanpunt, J., Shan, W.-X., and Tyler, B.M. (2001) High frequency mitotic gene conversion in genetic hybrids of the oomycete *Phytophthora sojae*. *Proceedings of the National Academy of Sciences USA* 98, 14530–14535. https://doi.org/10.1073/pnas.251464498

Chaparro, J.M., Sheflin, A.M., Manter, D.K., and Vivanco, J.M. (2012) Manipulating the soil microbiome to increase soil health and plant fertility. *Biology and Fertility of Soils* 48, 489–499. https://doi.org/10.1007/s00374-012-0691-4

Chaube, H.S. and Singh, U.S. (1991) *Plant Disease Management: Principles and Practices*. CRC Press, Boca Raton, Florida.

Chen, K. and Gao, C. (2014) Targeted genome modification technologies and their applications in crop improvements. *Plant Cell Reports* 33, 575–583. https://doi.org/10.1007/s00299-013-1539-6

Chisholm, S.T., Coaker, G., Day, B., and Staskawicz, B.J. (2006) Host–microbe interactions: shaping the evolution of the plant immune response. *Cell* 124, 803–814. https://doi.org/10.1016/j.cell.2006.02.008

Collard, B., Jahufer, M., Brouwer, J., and Pang, E. (2005) An introduction to markers, quantitative trait loci (QTL) mapping and marker-assisted selection for crop improvement: the basic concepts. *Euphytica* 142, 169–196. https://doi.org/10.1007/s10681-005-1681-5

Copping, L.G., and Menn, J.J. (2000) Biopesticides: a review of their action, applications and efficacy. *Pest Management Science* 56, 651–676. https://doi.org/10.1002/1526-4998(200008)56:8<651::AID-PS201>3.0.CO;2-U

Corrêa-Ferreira, B.S., and De Azevedo, J. (2002) Soybean seed damage by different species of stink bugs. *Agricultural and Forest Entomology* 4, 145–150. https://doi.org/10.1046/j.1461-9563.2002.00136.x

Dalton, T.J. (2004) A household hedonic model of rice traits: economic values from farmers in West Africa. *Agricultural Economics* 31, 149–159. https://doi.org/10.1111/j.1574-0862.2004.tb00253.x

Davis, R. and Tisdell, C.A. (2002) Alternative specifications and extensions of the economic threshold concept and the control of livestock pests. In: Hall, D.C. and Moffitt, L.J. (eds) *Economics of Pesticides, Sustainable Food Production, and Organic Food Markets.* Emerald Group Publishing, Bingley, UK, pp. 55–79.

Donatelli, M., Magarey, R.D., Bregaglio, S., Willocquet, L., Whish, J.P.M., and Savary, S. (2017) Modelling the impacts of pests and diseases on agricultural systems. *Agricultural Systems* 155, 213–224. https://doi.org/10.1016/j.agsy.2017.01.019

Dorrance, A.E. and Schmitthenner, A.F. (2000) New sources of resistance to *Phytophthora sojae* in the soybean plant introductions. *Plant Disease* 84, 1303–1308. https://doi.org/10.1094/PDIS.2000.84.12.1303

Dorrance, A., McClure, S., and DeSilva, A. (2003a) Pathogenic diversity of *Phytophthora sojae* in Ohio soybean fields. *Plant Disease* 87, 139–146. https://doi.org/10.1094/PDIS.2003.87.2.139

Dorrance, A.E., McClure, S.A., and St. Martin, S.K. (2003b) Effect of partial resistance on *Phytophthora* stem rot incidence and yield of soybean in Ohio. *Plant Disease* 87, 308–312. https://doi.org/10.1094/PDIS.2003.87.3.308

Dorrance, A.E., Berry, S., Bowen, P., and Lipps, P. (2004) Characterization of *Pythium* spp. from three Ohio fields for pathogenicity on corn and soybean and metalaxyl sensitivity. *Plant Health Progress,* https://doi.org/10.1094/PHP-2004-0202-01-RS

Dorrance, A., Robertson, A., Cianzo, S., Giesler, L., Grau, C., Draper, M., Tenuta, A.U. and Anderson, T. (2009) Integrated management strategies for *Phytophthora sojae* combining host resistance and seed treatments. *Plant Disease* 93, 875–882. https://doi.org/810.1094/PDIS-1093-1099-0875

Dorrance, A., Kurle, J., Robertson, A., Bradley, C., Giesler, L., Wise, K., and Concibido, V. (2016) Pathotype diversity of *Phytophthora sojae* in eleven states in the United States. *Plant Disease* 100, 1429–1437. https://doi.org/10.1094/PDIS-08-15-0879-RE

Doupnik, B. (1993) Soybean production and disease loss estimates for north central United States from 1989 to 1991. *Plant Disease* 77, 1170–1171. http://doi.org/10.1094/PD-77-1170

Drenth, A., Whisson, S., Maclean, D., Irwin, J., Obst, N., and Ryley, M. (1996) The evolution of races of *Phytophthora sojae* in Australia. *Phytopathology* 86, 163–169. https://doi.org/10.1094/Phyto-86-163

Drewes, M., Tietjen, K., and Sparks, T.C. (2012) High-throughput screening in agrochemical research. In: Jeschke, P., Kramer, W., Schirmer, U., and Witschel, M. (eds) *Modern Methods in Crop Protection Research.* Wiley, Weinheim, Germany, pp. 3–20.

DuPont Pioneer (2016) DuPont Pioneer announces intentions to commercialize first CRISPR-Cas product. Available at: www.pioneer.com/home/site/about/news-media/news-releases/template.CONTENT/guid.1DB8FB71-1117-9A56-E0B6-3EA6F85AAE92 (accessed 23 March 2018).

Elliot, S. (2016) Digital plant diagnosis: turning a mobile app into an agricultural game-changer. USDA National Institute of Food and Agriculture. Available at: https://nifa.usda.gov/blog/digital-plant-diagnosis-turning-mobile-app-agricultural-game-changer (accessed 2 December 2016).

Ellis, M.L., Broders, K.D., Paul, P.A., and Dorrance, A.E. (2010) Infection of soybean seed by *Fusarium graminearum* and effect of seed treatments on disease under controlled conditions. *Plant Disease* 95, 401–407. https://doi.org/10.1094/PDIS-05-10-0317

Endo, M., Mikami, M., and Toki, S. (2015) Multigene knockout utilizing off-target mutations of the CRISPR/Cas9 system in rice. *Plant and Cell Physiology* 56, 41–47. https://doi.org/10.1093/pcp/pcu154

Erwin, N. (2016) Data farming: how Big Data is revolutionizing Big Ag. Available at: http://ohiovalleyresource.org/2016/09/16/data-farming-big-data-revolutionizing-big-ag/(accessed 19 January 2017).

Espinosa, J.A. and Goodwin, B.K. (1991) Hedonic price estimation for Kansas wheat characteristics. *Western Journal of Agricultural Economics* 16, 72–85.

Falck-Zepeda, J. and Cohen, J.I. (2006) Biosafety regulation of genetically modified orphan crops in developing countries: a way forward. In: Just, R.E., Alston, J.M., and Zilberman, D. (eds) *Regulating Agricultural Biotechnology: Economics and Policy*. Springer, Boston, Massachusetts, pp. 509–533.

Fawke, S., Doumane, M., and Schornack, S. (2015) Oomycete interactions with plants: infection strategies and resistance principles. *Microbiology and Molecular Biology Reviews* 79, 263–280. https://doi.org/10.1128/MMBR.00010-15

Feder, G. (1979) Pesticides, Information, and pest management under uncertainty. *American Journal of Agricultural Economics* 61, 97–103. https://doi.org/10.2307/1239507

Fisher, R.A. (1936) Has Mendel's work been rediscovered? *Annals of Science* 1, 115–137. https://doi.org/10.1080/00033793600200111

Förster, H., Tyler, B.M., and Coffey, M.D. (1994) *Phytophthora sojae* races have arisen by clonal evolution and by rare outcrosses. *Molecular Plant–Microbe Interactions* 7, 780–791. http://doi.org/10.1094/MPMI-7-0780

Freebairn, J.W. (1992) Evaluating the level and distribution of benefits from dairy industry research. *Australian Journal of Agricultural Economics* 36, 141–165. https://doi.org/10.1111/j.1467-8489.1992.tb00517.x

Frisvold, G.B. (1997) Multimarket effects of agricultural research with technological spillovers. In: Hertel, T.W. (ed.) *Global Trade Analysis: Modeling and Applications*. Cambridge University Press, New York , pp. 321–346.

Fry, W.E. (1982) *Principles of Plant Disease Management*. Academic Press, New York.

Fry, W.E. (2016) *Phytophthora infestans*: new tools (and old ones) lead to new understanding and precision management. *Annual Review of Phytopathology* 54, 529–547. https://doi.org/10.1146/annurev-phyto-080615-095951

Fry, W.E. and Goodwin, S.B. (1997a) Re-emergence of potato and tomato late blight in the United States. *Plant Disease* 81, 1349–1357. https://dx.doi.org/10.1094/PDIS.1997.81.12.1349

Fry, W.E. and Goodwin, S.B. (1997b) Resurgence of the Irish potato famine fungus. *BioScience* 47, 363–371. https://doi.org/10.2307/1313151

Fuchs, M. and Gonslaves, D. (1997) Genetic engineering. In: Rechcigl, N.A. and Rechcigl, J.E. (eds) *Environmentally Safe Approaches to Crop Disease Control*. CRC Press, Boca Raton, Florida, pp. 333–368.

Garbeva, P., van Veen, J.A., and van Elsas, J.D. (2004) Microbial diversity in soil: selection of microbial populations by plant and soil type and implications for disease suppressiveness. *Annual Review of Phytopathology* 42, 243–270. https://doi.org/10.1146/annurev.phyto.42.012604.135455

Garrett, R.D., Lambin, E.F., and Naylor, R.L. (2013) Land institutions and supply chain configurations as determinants of soybean planted area and yields in Brazil. *Land Use Policy* 31, 385–396. https://doi.org/10.1016/j.landusepol.2012.08.002

Gent, D.H., Mahaffee, W.F., McRoberts, N., and Pfender, W.F. (2013) The use and role of predictive systems in disease management. *Annual Review of Phytopathology* 51, 267–289. https://doi.org/10.1146/annurev-phyto-082712-102356

Ghorbani, R., Wilcockson, S., Koocheki, A., and Leifert, C. (2008) Soil management for sustainable crop disease control: a review. *Environmental Chemistry Letters* 6, 149–162. https://doi.org/10.1007/s10311-008-0147-0

Gill, C. (2016) Artificial intelligence could help farmers diagnose crop diseases. Penn State News, 4 October, 2016. Available at: https://news.psu.edu/story/429727/2016/10/04/research/artificial-intelligence-could-help-farmers-diagnose-crop-diseases (accessed 5 January 2017).

Gill, U.S., Lee, S., and Mysore, K.S. (2015) Host versus nonhost resistance: distinct wars with similar arsenals. *Phytopathology* 105, 580–587. https://doi.org/10.1094/PHYTO-11-14-0298-RVW

Gleick, J. (1987) *Chaos: Making a New Science*. Penguin Books, New York.

Gobbato, E., Wang, E., Higgins, G., Bano, S.A., Henry, C., Schultze, M., and Oldroyd, G.E. (2013) *RAM1* and *RAM2* function and expression during arbuscular mycorrhizal symbiosis and *Aphanomyces euteiches* colonization. *Plant Signaling and Behavior* 8, e26049. https://doi.org/10.4161/psb.26049

Gordon, S.G., St. Martin, S.K., and Dorrance, A.E. (2006) *Rps8* maps to a resistance gene rich region on soybean molecular linkage group F. *Crop Science* 46, 168–173. https://doi.org/10.2135/cropsci2004.04-0024

Gould, A.B. and Hillman, B.I. (1997) Sanitation, eradication, exclusion, and quarantine. In: Rechcigl, N.A. and Rechcigl, J.E. (eds) *Environmentally Safe Approaches to Crop Disease Control*. CRC Press, Boca Raton, Florida, pp. 223–241.

Grau, C.R., Dorrance, A.E., Bond, J., and Russin, J.S. (2004) Fungal diseases. In: Boerma, H.R. and Specht, J.E. (eds) *Soybeans: Improvement, Production, and Uses*. American Society of Agronomy, Madison, Wisconsin.

Grijalba, P.E. and Gally, M.E. (2015) Virulence of *Phytophthora sojae* in the Pampeana Subregion of Argentina from 1998 to 2004. *Journal of Phytopathology* 163, 723–730. https://doi.org/10.1111/jph.12369

Griliches, Z. (1958) Research costs and social returns: hybrid corn and related innovations. *Journal of Political Economy* 66, 419. https://doi.org/10.1086/258077

Guo, J., Wang, Y., Song, C., Zhou, J., Qiu, L., Huang, H., and Wang, Y. (2010) A single origin and moderate bottleneck during domestication of soybean (*Glycine max*): implications from microsatellites and nucleotide sequences. *Annals of Botany* 106, 505–514. https://doi.org/10.1093/aob/mcq125

Hahnemann, W.M. (1991) Willingness to pay and willingness to accept: how much can they differ? *American Economic Review* 81, 635–647. https://doi.org/10.1257/000282803321455430

Hall, D.C. and Norgaard, R.B. (1973) On the timing and application of pesticides. *American Journal of Agricultural Economics* 55, 198–201. https://doi.org/10.2307/1238437

Hamlen, R.A., Labit, B., Bruhn, J.A., Holliday, M.J., Smith, C.M., and Shillingford, C.A. (1997) Use of chemical measures. In: Rechcigl, N.A. and Rechcigl, J.E. (eds) *Environmentally Safe Approaches to Crop Disease Control*. CRC Press, Boca Raton, Florida, pp. 65–92.

Harper, C.C. (1958) *Missouri and the Soybean Processing Industry*. Missouri Division of Resources and Development, Jefferson City, Missouri.

Hartman, G.L., Sinclair, J.B., and Rupe, J.C. (eds). (1999) *Compendium of Soybean Diseases*, 4th edn. American Phytopathological Society, St. Paul, Minnesota.

Hartung, F. and Schiemann, J. (2014) Precise plant breeding using new genome editing techniques: opportunities, safety and regulation in the EU. *The Plant Journal* 78, 742–752. https://doi.org/10.1111/tpj.12413

Headley, J. (1972) Defining the economic threshold. In: *Pest Control Strategies for the Future*. National Academy Of Sciences, Washington, DC, pp. 100–108.

Helmberger, P.G. and Chavas, J.-P. (1996) *The Economics of Agricultural Prices*. Prentice-Hall, Upper Saddle River, New Jersey.

Ho, P.-T. (1975) *The Cradle of the East*. Chinese University of Hong Kong, Hong Kong/University of Chicago Press, Chicago, Illinois.

Holm, R.E. and Baron, J.J. (2002) Evolution of the crop protection industry. In: Wheeler, W.B. (ed.) *Pesticides in Agriculture and the Environment*. Marcel Dekker, New York, pp. 295–326.

Hubbell, B.J., Marra, M.C., and Carlson, G.A. (2000) Estimating the demand for a new technology: Bt cotton and insecticide policies. *American Journal of Agricultural Economics* 82, 118–132. https://doi.org/10.1111/0002-9092.00010

Hudson, D. and Hite, D. (2003) Producer willingness to pay for precision application technology: implications for government and the technology industry. *Canadian Journal of Agricultural Economics* 51, 39–53. https://doi.org/10.1111/j.1744-7976.2003.tb00163.x

Hyde, J., Martin, M.A., Preckel, P.V., and Edwards, C.R. (1999) The economics of Bt corn: valuing protection from the European corn borer. *Review of Agricultural Economics* 21, 442–454. https://doi.org/10.2307/1349890

Hyde, J., Martin, M.A., Preckel, P.V., Buschman, L.L., Edwards, C.R., Sloderbeck, P.E., and Higgins, R.A. (2003) The value of Bt corn in southwest Kansas: a Monte Carlo simulation approach. *Journal of Agricultural and Resource Economics* 28, 15–33.

Hymowitz, T. (1970) On the domestication of the soybean. *Economic Botany* 24, 408–421. https://doi.org/10.1007/BF02860745

Hymowitz, T. (1984) Dorsett–Morse soybean collection trip to East Asia: 50 year retrospective. *Economic Botany* 38, 378–388. https://doi.org/10.1007/BF02859075

Hymowitz, T. (1990) Soybeans: the success story. In: Janick, J. and Simon, J. (eds) *Advances in New Crops*. Timber Press, Portland, Oregon, pp. 159–163.

Hymowitz, T. and Shurtleff, W. (2005) Debunking soybean myths and legends in the historical and popular literature. *Crop Science* 45, 473–476. https://doi.org/10.2135/cropsci2005.0473

ISAAA (2016) *Global Status of Commercialized Biotech/GM Crops: 2016.* ISAAA Brief No. 52. International Service for the Acquisition of Agri-biotech Applications, Ithaca, New York.

ISAAA (2018) GM Approval Database. Available at: www.isaaa.org/gmapprovaldatabase/default.asp (accessed 3 March 2018).

Janvier, C., Villeneuve, F., Alabouvette, C., Edel-Hermann, V., Mateille, T., and Steinberg, C. (2007) Soil health through soil disease suppression: which strategy from descriptors to indicators? *Soil Biology and Biochemistry* 39, 1–23. https://doi.org/10.1016/j.soilbio.2006.07.001

Jones, H.D. (2015) Regulatory uncertainty over genome editing. *Nature Plants* 1, 14011. https://doi.org/10.1038/nplants.2014.11

Jones, J.D.G. and Dangl, J.L. (2006) The plant immune system. *Nature* 444, 323. https://doi.org/10.1038/nature05286

Kahneman, D. and Tversky, A. (1979) Prospect theory: an analysis of decision under risk. *Econometrica* 47, 263–291. https://doi.org/10.2307/1914185

Kaitany, R.C., Hart, L.P., and Safir, G.R. (2001) Virulence composition of *Phytophthora sojae* in Michigan. *Plant Disease* 85, 1103–1106. https://doi.org/10.1094/PDIS.2001.85.10.1103

Kalaitzandonakes, N. and Bjornson, B. (1997) Vertical and horizontal coordination in the agro-biotechnology industry: evidence and implications. *Journal of Agricultural and Applied Economics* 29, 129–139. https://doi.org/10.1017/S1074070800029187

Kalaitzandonakes, N., and Kaufman, J. (2013) *Management of Early and Mid-season Soybean Diseases Caused by Oomycetes: Summary of 2013 Farmer Survey.* Working Paper 2013-8. Economics and Management of Agrobiotechnology Center, Columbia, Missouri.

Kalaitzandonakes, N. and Zahringer, K.A. (2018) Structural change and innovation in the global agricultural input sector. In: Kalaitzandonakes, N., Carayannis, E.G., Grigoroudis, A., and Rozakis, S. (eds) *From Agriscience to Agribusiness: Theories, Policies, and Practices in Technology Transfer and Commercialization.* Springer, Cham, Switzerland, pp. 75–99.

Kalaitzandonakes, N., Alston, J.M., and Bradford, K.J. (2007) Compliance costs for regulatory approval of new biotech crops. *Nature Biotechnology* 25, 509–511. https://doi.org/10.1038/nbt0507-509

Kalaitzandonakes, N., Miller, D.J., and Magnier, A. (2010) A worrisome crop? Is there market power in the US seed industry? *Regulation* 33, 20–26.

Kalaitzandonakes, N., Magnier, A., and Kolympiris, C. (2018) Innovation and technology transfer among firms in the agricultural input sector. In: Kalaitzandonakes, N., Carayannis, E.G., Grigoroudis, A., and Rozakis, S. (eds) *From Agriscience to Agribusiness: Theories, Policies, and Practices in Technology Transfer and Commercialization.* Springer, Cham, Switzerland, pp. 169–188.

Kamoun, S., Furzer, O., Jones, J.D., Judelson, H.S., Ali, G.S., Dalio, R.J., Roy, S.G., Schena, L., Zambounis, A., Panabières, F., *et al.* (2015) The top 10 oomycete pathogens in molecular plant pathology. *Molecular Plant Pathology* 16, 413–434. https://doi.org/10.1111/mpp.12190

Kashyap, P.L., Kumar, S., and Srivastava, A.K. (2016) Nanodiagnostics for plant pathogens. *Environmental Chemistry Letters* 15, 7–13. https://doi.org/10.1007/s10311-016-0580-4

Kaufmann, M.J. and Gerdemann, J.W. (1958) Root and stem rot of soybean caused by *Phytophthora sojae* n. sp. *Phytopathology* 48, 201–208. https://doi.org/10.1094/PHI-I-2007-0830-07

Keren, I.N., Menalled, F.D., Weaver, D.K., and Robison-Cox, J.F. (2015) Interacting agricultural pests and their effect on crop yield: application of a bayesian decision theory approach to the joint management of *Bromus tectorum* and *Cephus cinctus. PLoS One* 10, e0118111. https://doi.org/10.1371/journal.pone.0118111

Kilen, T.C., Hartwig, E.E., and Keeling, B.L. (1974) Inheritance of a second major gene for resistance to Phytophthora rot in soybeans. *Crop Science* 14, 260–262. https://doi.org/10.2135/cropsci1974.0011183X001400020027x

Kim, D.-S., Chun, S.-J., Jeon, J.-J., Lee, S.-W., and Joe, G.-H. (2004) Synthesis and fungicidal activity of ethaboxam against Oomycetes. *Pest Management Science* 60, 1007–1012. https://doi.org/10.1002/ps.873

Kim, H., Kim, S.-T., Ryu, J., Kang, B.-C., Kim, J.-S., and Kim, S.-G. (2017) CRISPR/Cpf1-mediated DNA-free plant genome editing. *Nature Communications* 8, 14406. https://doi.org/10.1038/ncomms14406

Kinkel, L.L., Bakker, M.G., and Schlatter, D.C. (2011) A coevolutionary framework for managing disease-suppressive soils. *Annual Review of Phytopathology* 49, 47–67. https://doi.org/10.1146/annurev-phyto-072910-095232

Kirkpatrick, M.T., Rupe, J.C., and Rothrock, C.S. (2006) Soybean response to flooded soil conditions and the association with soilborne plant pathogenic genera. *Plant Disease* 90, 592–596. https://doi.org/10.1094/PD-90-0592

Koenning, S.R. and Wrather, J.A. (2010) Suppression of soybean yield potential in the continental United States by plant diseases from 2006 to 2009. *Plant Health Progress,* https://doi.org/10.1094/PHP-2010-1122-01-RS

Kogan, M. (1994) Plant resistance in pest management. In: Metcalf, R.L. and Luckman, W.H. (eds) *Introduction to Pest Management,* 3rd edn. Wiley, New York, pp. 73–128.

Kogan, M. and Turnipseed, S. (1987) Ecology and management of soybean arthropods. *Annual Review of Entomology* 32, 507–538. https://doi.org/10.1146/annurev.en.32.010187.002451

Kristofersson, D. and Rickertsen, K. (2004) Efficient estimation of hedonic inverse input demand systems. *American Journal of Agricultural Economics* 86, 1127–1137. https://doi.org/10.1111/j.0002-9092.2004.00658.x

Krupinsky, J.M., Bailey, K.L., McMullen, M.P., Gossen, B.D., and Turkington, T.K. (2002) Managing plant disease risk in diversified cropping systems. *Agronomy Journal* 94, 198–209. https://doi.org/10.2134/agronj2002.1980

Lacey, L.A., Grzywacz, D., Shapiro-Ilan, D.I., Frutos, R., Brownbridge, M., and Goettel, M.S. (2015) Insect pathogens as biological control agents: back to the future. *Journal of Invertebrate Pathology* 132, 1–41. https://doi.org/10.1016/j.jip.2015.07.009

Lamberth, C., Jeanmart, S., Luksch, T., and Plant, A. (2013) Current challenges and trends in the discovery of agrochemicals. *Science* 341, 742–746.

Lancaster, K.J. (1966) A new approach to consumer theory. *Journal of Political Economy* 74, 132–157. https://doi.org/10.1086/259131

Langenbach, C., Campe, R., Beyer, S.F., Mueller, A.N., and Conrath, U. (2016) Fighting Asian soybean rust. *Frontiers in Plant Science* 7, 797. https://doi.org/10.3389/fpls.2016.00797

Leitz, R., Hartman, G., Pedersen, W., and Nickell, C. (2000) Races of *Phytophthora sojae* on soybean in Illinois. *Plant Disease* 84, 487. https://doi.org/10.1094/PDIS.2000.84.4.487D

Leskey, T.C., Hamilton, G.C., Nielsen, A.L., Polk, D.F., Rodriguez-Saona, C., Bergh, J.C., Herbert, D.A., Kuhar, T.P., Pfeiffer, D., Dively, G.P., *et al.* (2012) Pest status of the brown marmorated stink bug, *Halyomorpha halys*, in the USA. *Outlooks on Pest Management* 23, 218–226. https://doi.org/10.1564/23oct07

Levesque, C.A. and de Cock, A.W. (2004) Molecular phylogeny and taxonomy of the genus *Pythium*. *Mycological Research* 108, 1363–1383. https://doi.org/10.1017/S0953756204001431

Li, T., Liu, B., Spalding, M.H., Weeks, D.P., and Yang, B. (2012) High-efficiency TALEN-based gene editing produces disease-resistant rice. *Nature Biotechnology* 30, 390–392. https://doi.org/10.1038/nbt.2199

Lichtenberg, E., and Zilberman, D. (1986) The econometrics of damage control: why specification matters. *American Journal of Agricultural Economics* 68, 261–273. https://doi.org/10.2307/1241427

Lin, F., Zhao, M., Ping, J., Johnson, A., Zhang, B., Abney, T.S., Hughes, T.J., and Ma, J. (2013) Molecular mapping of two genes conferring resistance to *Phytophthora sojae* in a soybean landrace PI 567139B. *Theoretical and Applied Genetics* 126, 2177–2185. https://doi.org/10.1007/s00122-013-2127-4

Liu, D., Hu, R., Palla, K.J., Tuskan, G.A., and Yang, X. (2016) Advances and perspectives on the use of CRISPR/Cas9 systems in plant genomics research. *Current Opinion in Plant Biology* 30, 70–77. https://doi.org/10.1016/j.pbi.2016.01.007

Loso, M.R., Garizi, N., Hegde, V.B., Hunter, J.E., and Sparks, T.C. (2016) Lead generation in crop protection research: a portfolio approach to agrochemical discovery. *Pest Management Science* 73, 678–685. https://doi.org/10.1002/ps.4336

Luck, J., Spackman, M., Freeman, A., Griffiths, W., Finlay, K., and Chakraborty, S. (2011) Climate change and diseases of food crops. *Plant Pathology* 60, 113–121. https://doi.org/10.1111/j.1365-3059.2010.02414.x

Lusk, J.L. and Hudson, D. (2004) Willingness-to-pay estimates and their relevance to agribusiness decision making. *Applied Economic Perspectives and Policy* 26, 152–169. https://doi.org/10.1111/j.1467-9353.2004.00168.x

Lusser, M., Parisi, C., Plan, D., and Rodríguez-Cerezo, E. (2012) Deployment of new biotechnologies in plant breeding. *Nature Biotechnology* 30, 231–239. https://doi.org/10.1038/nbt.2142

Madden, L.V. and Hughes, G. (1995) Plant disease incidence: distributions, heterogeneity, and temporal analysis. *Annual Review of Phytopathology* 33, 529–564. https://doi.org/10.1146/annurev.py.33.090195.002525

Magarey, R.D., Colunga-Garcia, M., and Fieselmann, D.A. (2009) Plant biosecurity in the United States: roles, responsibilities, and information needs. *BioScience* 59, 875–884. https://doi.org/10.1525/bio.2009.59.10.9

Mahlein, A.-K. (2015) Plant disease detection by imaging sensors – parallels and specific demands for precision agriculture and plant phenotyping. *Plant Disease* 100, 241–251. https://doi.org/10.1094/PDIS-03-15-0340-FE

Malvick, D. and Grunden, E. (2004) Traits of soybean-infecting *Phytophthora* populations from Illinois agricultural fields. *Plant Disease* 88, 1139–1145. https://dx.doi.org/10.1094/PDIS.2004.88.10.1139

Marra, M.C., Rejesus, R.M., Roberts, R.K., English, B.C., Larson, J.A., Larkin, S.L., and Martin, S. (2010) Estimating the demand and willingness-to-pay for cotton yield monitors. *Precision Agriculture* 11, 215–238. https://doi.org/10.1007/s11119-009-9127-z

Martha, G.B., Contini, E., and Alves, E. (2012) Embrapa: its origins and changes. In: Baer, W. (ed.) *The Regional Impact of National Policies: the Case of Brazil*. Edward Elgar, Cheltenham, UK, pp. 201–226.

Matthiesen, R., Ahmad, A., and Robertson, A. (2016) Temperature affects aggressiveness and fungicide sensitivity of four *Pythium* spp. that cause soybean and corn damping off in Iowa. *Plant Disease* 100, 583–591. https://doi.org/10.1094/PDIS-04-15-0487-RE

Matuschke, I., Mishra, R.R., and Qaim, M. (2007) Adoption and impact of hybrid wheat in India. *World Development* 35, 1422–1435. https://doi.org/10.1016/j.worlddev.2007.04.005

McRoberts, N., Hall, C., Madden, L.V., and Hughes, G. (2011) Perceptions of disease risk: from social construction of subjective judgments to rational decision making. *Phytopathology* 101, 654–665. https://doi.org/10.1094/PHYTO-04-10-0126

Menn, J.J. (1980) Contemporary frontiers in chemical pesticide research. *Journal of Agricultural and Food Chemistry* 28, 2–8. https://doi.org/10.1021/jf60227a012

Metcalf, R.L. (1980) Changing role of insecticides in crop protection. *Annual Review of Entomology* 25, 219–256. https://doi.org/10.1146/annurev.en.25.010180.001251

Moffitt, L.J. (1986) Risk-efficient thresholds for pest control decisions. *Journal of Agricultural Economics* 37, 69–75. https://doi.org/10.1111/j.1477-9552.1986.tb00318.x

Morton, V. and Staub, T. (2008) A short history of fungicides. American Phytopathological Society. Available at: www.apsnet.org/publications/apsnetfeatures/Pages/Fungicides.aspx (accessed 16 October 2017).

Mueller, D.S., Wise, K.A., Dufault, N.S., Bradley, C.A., and Chilvers, M.I. (eds). (2013) *Fungicides for Field Crops*. American Phytopathological Society, St. Paul, Minnesota.

Mueller, E.H., Athow, K.L., and Laviolette, F.A. (1978) Inheritance of resistance to four physiologic races of *Phytophthora megasperma* var. *sojae*. *Phytopathology* 68, 1318–1322. https://doi.org/10.1094/Phyto-68-1318

Mulla, D.J. (2013) Twenty five years of remote sensing in precision agriculture: key advances and remaining knowledge gaps. *Biosystems Engineering* 114, 358–371. https://doi.org/10.1016/j.biosystemseng.2012.08.009

Murillo-Williams, A. and Pedersen, P. (2008) Early incidence of soybean seedling pathogens in Iowa. *Agronomy Journal* 100, 1481–1487. https://doi.org/10.2134/agronj2007.0346

Naab, J., Singh, P., Boote, K., Jones, J., and Marfo, K. (2004) Using the CROPGRO-peanut model to quantify yield gaps of peanut in the Guinean Savanna Zone of Ghana. *Agronomy Journal* 96, 1231–1242. https://doi.org/10.2134/agronj2004.1231

NASS (2012) Agricultural Resource Management Survey: U.S. soybean industry. USDA National Agricultural Statistics Service. Available at: www.nass.usda.gov/Surveys/Guide_to_NASS_Surveys/Ag_Resource_Management/ARMS_Soybeans_Factsheet/ (accessed 21 September 2018).

National Research Council (2000) *The Future Role of Pesticides in US Agriculture*. National Academies Press, Washington, DC.

Nelson, B. (1999a) Fusarium blight or wilt, root rot, and pod and collar rot. In: Hartman, G.L., Sinclair, J.B., and Rupe, J.C. (eds) *Compendium of Soybean Diseases*, 4th edn. American Phytopathological Society Press, St. Paul, Minnesota, pp. 35–36.

Nelson, B. (1999b) *Fusarium* spp. associated with seeds. In: Hartman, G.L., Sinclair, J.B., and Rupe, J.C. (eds) *Compendium of Soybean Diseases*, 4th edn. American Phytopathological Society Press, St. Paul, Minnesota, p. 37.

Nelson, B.D., Mallik, I., McEwen, D., and Christianson, T. (2008) Pathotypes, distribution, and metalaxyl sensitivity of *Phytophthora sojae* from North Dakota. *Plant Disease* 92, 1062–1066. https://doi.org/10.1094/PDIS-92-7-1062

Niblack, T.L., Tylka, G.L., and Riggs, R.D. (2004) Nematode pathogens of soybean. In: Boerma, H.R. and Specht, J.E. (eds) *Soybeans: Improvement, Production, and Uses*. American Society of Agronomy, Madison, Wisconsin, pp. 821–851.

Noel, G.R. (1999) Diseases caused by nematodes. In: Hartman, G.L., Sinclair, J.B., and Rupe, J.C. (eds) *Compendium of Soybean Diseases*, 4th edn. American Phytopathological Society, St. Paul, Minnesota, pp. 50–51.

Norris, P.E., and Kramer, R.A. (1990) The elicitation of subjective probabilities with applications in agricultural economics. *Review of Marketing and Agricultural Economics* 58, 127–147.

Nutter, F.W., Teng, P.S., and Royer, M.H. (1993) Terms and concepts for yield, crop loss, and disease thresholds. *Plant Disease* 77, 211–215. http://www.apsnet.org/publications/ plantdisease/backissues/Documents/1993Articles/PlantDisease77n02_211.PDF

NY DEC (2015) *Active Ingredient Data Package: Metalaxyl and Mefenoxam*. New York Department of Environmental Conservation. Available at: www.dec.ny.gov/docs/materials_ minerals_pdf/mefenoxamdata.pdf (accessed 21 September 2018).

Oerke, E.-C. (1999) Estimated crop losses due to pathogens, animal pests, and weeds. In: Oerke, E.-C., Dehne, H.-W., Schoenbeck, F., and Weber, A. (eds) *Crop Production and Crop Protection: Estimated Losses in Major Food and Cash Crops*. Elsevier Science, Amsterdam, The Netherlands, pp. 72–741.

Oerke, E.-C. (2006) Crop losses to pests. *Journal of Agricultural Science* 144, 31–43. https:// doi.org/10.1017/S0021859605005708

Oerke, E.-C. and Dehne, H.-W. (2004) Safeguarding production - losses in major crops and the role of crop protection. *Crop Protection* 23, 275–285. https://doi.org/10.1016/j. cropro.2003.10.001

Osborne, S.L., and Riedell, W.E. (2006) Starter nitrogen fertilizer impact on soybean yield and quality in the Northern Great Plains. *Agronomy Journal* 98, 1569–1574. https://doi. org/10.2134/agronj2006.0089

Padmanabhan, S. (1973) The great Bengal famine. *Annual Review of Phytopathology* 11, 11–24. https://doi.org/10.1146/annurev.py.11.090173.000303

Pannell, D.J. (1990) Responses to risk in weed control decisions under expected profit maximisation.*Journal of Agricultural Economics* 41, 391–401. https://doi.org/10.1111/j.1477-9552.1990. tb00655.x

Pedigo, L.P. and Rice, M.E. (2009) *Entomology and Pest Management*. Waveland Press, Long Grove, Illinois.

Phillips McDougall (2011) *The Cost and Time Involved in the Discovery, Development and Authorisation of a New Plant Biotechnology Derived Trait*. Crop Life International, Brussels, Belgium. Available at: http://croplife.org/wp-content/uploads/pdf_files/Getting-a-Biotech-Crop-to-Market-Phillips-McDougall-Study.pdf (accessed 21 September 2018).

Phillips McDougall (2016a) *Agriservice Report: Products Section, 2015 Market*. Phillips McDougall, Pathhead, UK.

Phillips McDougall (2016b) *The Cost of New Agrochemical Product Discovery, Development and Registration in 1995, 2000, 2005–8 and 2010 to 2014*. CropLife International, Brussels, Belgium. Available at: http://191hmt1pr08amfq62276etw2.wpengine.netdna-cdn.com/ wp-content/uploads/2016/04/Phillips-McDougall-Final-Report_4.6.16.pdf (accessed 21 September 2018).

Pindyck, R.S. and Rubinfeld, D.L. (1981) *Econometric Models and Economic Forecasts*. McGraw-Hill, New York.

Pingali, P. and Carlson, G.A. (1985) Human capital, adjustments in subjective probabilities, and the demand for pest controls. *American Journal of Agricultural Economics* 67, 853–861. https://doi.org/10.2307/1241826

Plant, R.E. (1986) Uncertainty and the economic threshold. *Journal of Economic Entomology* 79, 1–6. https://doi.org/10.1093/jee/79.1.1

Ploper, L.D., Athow, K.L., and Laviolette, F.A. (1985) A new allele at *Rps3* locus for resistance to *Phytophthora megasperma* f. sp. *glycinea* in soybean. *Phytopathology* 75, 690–694. https://doi.org/10.1094/Phyto-75-690

Powles, S.B. (2008) Evolved glyphosate-resistant weeds around the world: lessons to be learnt. *Pest Management Science* 64, 360–365. https://doi.org/10.1002/ps.1525

Praneetvatakul, S., Kuwattanasiri, D., and Waibel, H. (2002) The productivity of pesticide use in rice production of Thailand: a damage control approach. Paper presented at the *International Symposium on Sustaining Food Security and Managing Natural Resources in Southeast Asia: Challenges for the 21st Century*, 8–11 January 2002, Chiang Mai, Thailand.

PVPO (2017) Plant Variety Protection Office – scanned certificates. USDA Agricultural Marketing Service. Available at: https://apps.ams.usda.gov/CMS/ (accessed 24 September 2018).

Qaim, M. and de Janvry, A. (2003) Genetically modified crops, corporate pricing strategies, and farmers' adoption: the case of Bt cotton in Argentina. *American Journal of Agricultural Economics* 85, 814–828. https://doi.org/10.1111/1467-8276.00490

Qiu, L.-J. and Chang, R.-Z. (2010) The origin and history of soybean. In: Singh, G. (ed.) *The Soybean: Botany, Production and Uses*. CAB International, Wallingford, UK, pp. 1–23.

Radmer, L., Anderson, G., Malvick, D., Kurle, J., Rendahl, A., and Mallik, A. (2017) *Pythium*, *Phytophthora*, and *Phytopythium* spp. associated with soybean in Minnesota, their relative aggressiveness on soybean and corn, and their sensitivity to seed treatment fungicides. *Plant Disease* 101, 62–72.

Ragsdale, D.W., Landis, D.A., Brodeur, J., Heimpel, G.E., and Desneux, N. (2011) Ecology and management of the soybean aphid in North America. *Annual Review of Entomology* 56, 375–399. https://doi.org/10.1094/PDIS-02-16-0196-RE

REEIS (2017) Integrated management of oomycete diseases of soybean and other crop plants. USDA Research, Education, and Economics Information System. Available at: https://reeis.usda.gov/web/crisprojectpages/0224426-integrated-management-of-oomycete-diseases-of-soybean-and-other-crop-plants.html (accessed 24 September 2018).

REM (2014) *Cost Increases Caused by Resistant and Tolerant Weeds*. AAPRESID, Buenos Aires, Argentina.

Retzinger, E.J. and Mallory-Smith, C. (1997) Classification of herbicides by site of action for weed resistance management strategies. *Weed Technology* 11, 384–393. https://doi.org/10.1614/0890-037X(2003)017[0605:RCOHBS]2.0.CO;2

Ridley, S.M., Elliott, A.C., Yeung, M., and Youle, D. (1998) High-throughput screening as a tool for agrochemical discovery: automated synthesis, compound input, assay design and process management. *Pest Management Science* 54, 327–337. https://doi.org/10.1002/(SICI)1096-9063(199812)54:4<327::AID-PS828>3.0.CO;2-C

Riggs, R.D. and Niblack, T.L. (1999) Soybean cyst nematode. In: Hartman, G.L., Sinclair, J.B., and Rupe, J.C. (eds) *Compendium of Soybean Diseases*, 4th edn. American Phytopathological Society, St. Paul, Minnesota, pp. 52–53.

Rizvi, S. and Yang, X.B. (1996) Fungi associated with soybean seedling disease in Iowa. *Plant Disease* 80, 57–60. https://doi.org/10.1094/PD-80-0057

Robertson, A., Cerra, S., and Cianzio, S. (2007) Populations of *Phytophthora sojae* are diverse within single fields. IOWA State University Extension and Outreach, Integrated Crop Management. Available at: www.ipm.iastate.edu/ipm/icm/2007/6-18/psojae.html (accessed 24 September 2018).

Rojas, A., Jacobs, J., Napieralski, S., Karaj, B., Bradley, C., Chase, T., Esker, P.D., Giesler, L.J., Jardine, D.J., Malvick, D.K., *et al.* (2017a) Oomycete species associated with soybean seedlings in North America – part 1: identification and pathogenicity characterization. *Phytopathology* 107, 280–292. https://doi.org/10.1094/PHYTO-04-16-0177-R

Rojas, A., Jacobs, J., Napieralski, S., Karaj, B., Bradley, C., Chase, T., Esker, P.D., Giesler, L.J., Jardine, D.J., Malvick, D.K., *et al.* (2017b) Oomycete species associated with soybean seedlings in North America – part 2: diversity and ecology in relation to environmental and edaphic factors. *Phytopathology* 107, 293–304. https://doi.org/10.1094/PHYTO-04-16-0176-R

Rosenzweig, C., Iglesias, A., Yang, X., Epstein, P.R., and Chivian, E. (2001) Climate change and extreme weather events – implications for food production, plant diseases, and pests. *Global Change and Human Health* 2, 90–104. https://doi.org/10.1023/A:1015086831467

Ruan, J., Li, H., Xu, K., Wu, T., Wei, J., Zhou, R., Liu, Z., Mu, Y., Yang, S., Ouyang, H., *et al.* (2015) Highly efficient CRISPR/Cas9-mediated transgene knockin at the H11 locus in pigs. *Scientific Reports* 5, 14253. https://doi.org/10.1038/srep14253

Rupe, J.C. and Hartman, G.L. (1999) Sudden death syndrome. In: Hartman, G.L., Sinclair, J.B., and Rupe, J.C. (eds) *Compendium of Soybean Diseases*, 4th edn. American Phytopathological Society, St. Paul, Minnesota, pp. 37–39.

Rush, C.M., Piccinni, G., and Harveson, R.M. (1997) Agronomic measures. In: Rechcigl, N.A. and Rechcigl, J.E. (eds) *Environmentally Safe Approaches to Crop Disease Control*. CRC Press, Boca Raton, Florida, pp. 234–282.

Ryley, M.J., Obst, N.R., Irwin, J.A.G., and Drenth, A. (1998) Changes in the racial composition of *Phytophthora sojae* in Australia between 1979 and 1996. *Plant Disease* 82, 1048–1054. https://doi.org/10.1094/PDIS.1998.82.9.1048

Saini, V. and Kumar, A. (2014) Computer aided pesticide design: a rational tool for supplementing DDT lacunae. *Chemical Science Transactions* 3, 676–688. https://doi.org/10.7598/cst2014.719

Sandhu, D., Gao, H., Cianzio, S., and Bhattacharyya, M.K. (2004) Deletion of a disease resistance nucleotide-binding-site leucine-rich repeat-like sequence is associated with the loss of the *Phytophthora* resistance gene *Rps4* in soybean. *Genetics* 168, 2157–2167. https://doi.org/10.1534/genetics.104.032037

Sanogo, S. and Yang, X.B. (2001) Relation of sand content, pH, and potassium and phosphorus nutrition to the development of sudden death syndrome in soybean. *Canadian Journal of Plant Pathology* 23, 174–180. https://doi.org/10.1080/07060660109506927

Scherkenbeck, J. (2009) Modern trends in agrochemistry. *Bioorganic and Medicinal Chemistry* 17, 4019. https://doi.org/10.1016/j.bmc.2009.05.050

Schettini, T., Legg, E., and Michelmore, R. (1991) Insensitivity to metalaxyl in California populations of *Bremia lactucae* and resistance of California lettuce cultivars to downy mildew. *Phytopathology* 81, 64–70. https://doi.org/10.1094/Phyto-81-64

Schlub, R.L. and Lockwood, J. (1981) Etiology and epidemiology of seedling rot of soybean by *Pythium ultimum*. *Phytopathology* 71, 134–138. https://doi.org/10.1094/Phyto-71-134

Schmitthenner, A.F. (1985) Problems and progress in control of *Phytophthora* root rot of soybean. *Plant Disease* 69, 362–368. https://doi.org/10.1094/PD-69-362

Schmitthenner, A.F. (1999) *Phytophthora* rot. In: Hartman, G.L., Sinclair, J.B., and Rupe, J.C. (eds) *Compendium of Soybean Diseases*, 4th edn. American Phytopathological Society, St. Paul, Minnesota, pp. 39–42.

Schmitthenner, A., Hobe, M., and Bhat, R. (1994) *Phytophthora sojae* races in Ohio over a 10-year interval. *Plant Disease* 78, 269–276. https://doi.org/10.1094/PD-78-0269

Schuh, W. (1999) *Cercospora* blight, leaf spot, and purple seed stain. In: Hartman, G.L., Sinclair, J.B., and Rupe, J.C. (eds) *Compendium of Soybean Diseases*, 4th edn. American Phytopathological Society, St. Paul, Minnesota, pp. 17–18.

Schumann, G.L. and D'Arcy, C.J. (2010) *Essential Plant Pathology*. American Phytopathological Society, St. Paul, Minnesota.

Servick, K. (2015) U.S. to review agricultural biotech regulations. *Science* 349, 131. https://doi.org/10.1126/science.349.6244.131

Sexton, S.E., Lei, Z., and Zilberman, D. (2007) The economics of pesticides and pest control. *International Review of Environmental and Resource Economics* 1, 271–326. https://doi.org/10.1561/101.00000007

Shea, K., Possingham, H.P., Murdoch, W.W., and Roush, R. (2002) Active adaptive management in insect pest and weed control: intervention with a plan for learning. *Ecological*

Applications 12, 927–936. https://doi.org/10.1890/1051-0761(2002)012[0927:AAMIIP] 2.0.CO;2

Shelton, K.A. and Lahm, G.P. (2015) Building a successful crop protection pipeline: molecular starting points for discovery. In: Maienfisch, P. and Stevenson, T.M. (eds) *Discovery and Synthesis of Crop Protection Products*. ACS Publications, Washington, DC, pp. 15–23.

Shukla, V.K., Doyon, Y., Miller, J.C., DeKelver, R.C., Moehle, E.A., Worden, S.E., Mitchell, J.C., Arnold, N.L., Gopalan, S., Meng, X., *et al.* (2009) Precise genome modification in the crop species *Zea mays* using zinc-finger nucleases. *Nature* 459, 437–441. https://doi. org/10.1038/nature07992

Sinclair, J.B. (1999) Bacterial diseases. In: Hartman, G.L., Sinclair, J.B., and Rupe, J.C. (eds) *Compendium of Soybean Diseases*, 4th edn. American Phytopathological Society, St. Paul, Minnesota, p. 5.

Sinclair, J.B., and Hartman, G.L. (1999a) Brown spot. In: Hartman, G.L., Sinclair, J.B., and Rupe, J.C. (eds) *Compendium of Soybean Diseases*, 4th edn. American Phytopathological Society, St. Paul, Minnesota, p. 16.

Sinclair, J.B., and Hartman, G.L. (1999b) Diseases caused by fungi. In: Hartman, G.L., Sinclair, J.B., and Rupe, J.C. (eds) *Compendium of Soybean Diseases*, 4th edn. American Phytopathological Society, St. Paul, Minnesota, pp. 11–12.

Sinclair, J.B. and Hartman, G.L. (1999c) Soybean rust. In: Hartman, G.L., Sinclair, J.B., and Rupe, J.C. (eds) *Compendium of Soybean Diseases*, 4th edn. American Phytopathological Society, St. Paul, Minnesota, pp. 25–26.

Smith, G.S. and Wyllie, T.D. (1999) Charcoal rot. In: Hartman, G.L., Sinclair, J.B., and Rupe, J.C. (eds) *Compendium of Soybean Diseases*, 4th edn. American Phytopathological Society, St. Paul, Minnesota, pp. 29–30.

Sonka, S.T., Bender, K.L., and Fisher, D.K. (2004) Economics and marketing. In: Boerma, H.R. and Specht, J.E. (eds) *Soybeans: Improvement, Production, and Uses*. American Society of Agronomy, Madison, Wisconsin, pp. 919–947.

Sparks, T.C. (2013) Insecticide discovery: an evaluation and analysis. *Pesticide Biochemistry and Physiology* 107, 8–17. https://doi.org/10.1016/j.pestbp.2013.05.012

Sparks, T.C. and Lorsbach, B.A. (2016) Perspectives on the agrochemical industry and agro-chemical discovery. *Pest Management Science* 73, 672–677. https://doi.org/10.1002/ps.4457

Sparks, T.C. and Nauen, R. (2015) IRAC: mode of action classification and insecticide resistance management. *Pesticide Biochemistry and Physiology* 121, 122–128. https://doi. org/10.1016/j.pestbp.2014.11.014

Speck-Planche, A., Natalia Dias Soeiro Cordeiro, M., Guilarte-Montero, L., and Yera-Bueno, R. (2011) Current computational approaches towards the rational design of new insecticidal agents. *Current Computer-Aided Drug Design* 7, 304–314. https://doi. org/10.2174/157340911798260359

Stern, V.M. (1973) Economic thresholds. *Annual Review of Entomology* 18, 259–280. https:// doi.org/10.1146/annurev.en.18.010173.001355

Stern, V.M., Smith, R.F., van den Bosch, R., and Hagen, K. (1959) The integration of chemical and biological control of the spotted alfalfa aphid: the integrated control concept. *Hilgardia* 29, 81–101. https://doi.org/10.3733/hilg.v29n02p081

Sternberg, S.H. and Doudna, J.A. (2015) Expanding the biologist's toolkit with CRISPR-Cas9. *Molecular Cell* 58, 568–574. https://doi.org/10.1016/j.molcel.2015.02.032

Stetter, J. (1993) Trends in the future development of pest and weed control – an industrial point of view. *Regulatory Toxicology and Pharmacology* 17, 346–370. https://doi. org/10.1006/rtph.1993.1035

Stewart, S., Abeysekara, N., and Robertson, A.E. (2014) Pathotype and genetic shifts in a population of *Phytophthora sojae* under soybean cultivar rotation. *Plant Disease* 98, 614–624. https://doi.org/10.1094/PDIS-05-13-0575-RE

Stine Seed Company (2016) Soybean breeding program – stine soybean. Available at: www. stineseed.com/research/soybean-breeding-program/ (accessed 2 August 2016).

Strange, R.N. and Scott, P.R. (2005) Plant disease: a threat to global food security. *Annual Review of Phytopathology* 43, 83–116. https://doi.org/10.1146/annurev.phyto.43.113004.133839

Sugimoto, T., Yoshida, S., Kaga, A., Hajika, M., Watanabe, K., Aino, M., Tatsuda, K., Yamamoto, R., Matoh, T., Walker, D.R., *et al.* (2011) Genetic analysis and identification of DNA markers linked to a novel *Phytophthora sojae* resistance gene in the Japanese soybean cultivar Waseshiroge. *Euphytica* 182, 133. https://doi.org/10.1007/s10681-011-0525-8

Sugimoto, T., Kato, M., Yoshida, S., Matsumoto, I., Kobayashi, T., Kaga, A., Hajika, M., Yamamoto, R., Watanabe, K., Aino, M., *et al.* (2012) Pathogenic diversity of *Phytophthora sojae* and breeding strategies to develop *Phytophthora*-resistant soybeans. *Breeding Science* 61, 511–522. https://doi.org/10.1270/jsbbs.61.511

Sun, P. and Yang, X. (2000) Light, temperature, and moisture effects on apothecium production of *Sclerotinia sclerotiorum*. *Plant Disease* 84, 1287–1293. https://doi.org/10.1094/PDIS.2000.84.12.1287

Sun, S., Wu, X., Zhao, J., Wang, Y., Tang, Q., Yu, D., Gai, J.Y., and Xing, H. (2011) Characterization and mapping of *RpsYu25*, a novel resistance gene to *Phytophthora sojae*. *Plant Breeding* 130, 139–143. https://doi.org/10.1111/j.1439-0523.2010.01794.x

Syngenta (2016a) Apron Maxx RTA seed treatment. Available at: www.syngentacropprotection. com/apron-maxx-rta-seed-treatment?tab=details (accessed 16 June 2016).

Syngenta (2016b) Apron XL seed treatment. Available at: www.syngentacropprotection.com/ apron-xl-seed-treatment?tab=details (accessed 16 June 2016).

Syngenta (2016c) Ridomil Gold SL fungicide. Available at: www.syngentacropprotection.com/ ridomil-gold-sl-fungicide?tab=details (accessed 16 June 2016).

Szmedra, P.I., Wetzstein, M.E., and McClendon, R.W. (1990) Economic threshold under risk: a case study of soybean production. *Journal of Economic Entomology* 83, 641–646. https:// doi.org/10.1093/jee/83.3.641

Takeshima, H. (2010) Prospects for development of genetically modified cassava in sub-Saharan Africa. *AgBioForum* 13, 63–75.

Tietjen, K. and Schreier, P.H. (2012) New targets for fungicides. In: Jeschke, P., Kramer, W., Schirmer, U., and Witschel, M. (eds) *Modern Methods in Crop Protection Research*. Wiley, Weinheim, Germany, pp. 197–216.

Tisdell, C.A. (1986) Levels of pest control and uncertainty of benefits. *Australian Journal of Agricultural and Resource Economics* 30, 157–161. https://doi.org/10.1111/j.1467-8489.1986. tb00663.x

Tolin, S.A. (1999) Diseases caused by viruses. In: Hartman, G.L., Sinclair, J.B., and Rupe, J.C. (eds) *Compendium of Soybean Diseases*, 4th edn. American Phytopathological Society, St. Paul, Minnesota, pp. 57–59.

Tolin, S.A. and Lacy, G.H. (2004) Viral, bacterial, and phytoplasmal diseases of soybean. In: Boerma, H.R. and Specht, J.E. (eds) *Soybeans: Improvement, Production, and Uses*. American Society of Agronomy, Madison, Wisconsin, pp. 765–820.

Trading Economics (2018) Soybeans. Available at: https://tradingeconomics.com/commodity/ soybeans (accessed 18 April 2018).

Turnipseed, S.G. and Kogan, M. (1976) Soybean entomology. *Annual Review of Entomology* 21, 247–282. https://doi.org/10.1146/annurev.en.21.010176.001335

Tyler, B.M. (2007) *Phytophthora sojae*: root rot pathogen of soybean and model oomycete. *Molecular Plant Pathology* 8, 1–8. https://doi.org/10.1111/j.1364-3703.2006.00373.x

Tyler, B.M., Tripathy, S., Zhang, X., Dehal, P., Jiang, R.H., Aerts, A., Arredondo, F.D., Baxter, L., Bensasson, D., Beynon, J.L., *et al.* (2006) *Phytophthora* genome sequences uncover evolutionary origins and mechanisms of pathogenesis. *Science* 313, 1261–1266. https:// doi.org/10.1126/science.1128796

Uchida, M., Roberson, R.W., Chun, S.-J., and Kim, D.-S. (2005) *In vivo* effects of the fungicide ethaboxam on microtubule integrity in *Phytophthora infestans*. *Pest Management Science* 61, 787–792. https://doi.org/10.1002/ps.1045

UCMP (2006) Introduction to the Oomycota. University of California Museum of Paleontology. Available at: www.ucmp.berkeley.edu/chromista/oomycota.html (accessed 15 July 2016).

UN (2015) World population prospects: the 2015 revision. United Nations, Department of Economic and Social Affairs, Population Division. Available at: http://esa.un.org/unpd/wpp/DataQuery/ (accessed 17 November 2015).

USDA (2017a) *Argentina Oilseeds and Products Annual Report*. GAIN Report. USDA Foreign Agriculture Service, Washington, DC.

USDA (2017b) *India Oilseeds and Products Annual Report*. GAIN Report IN7039. USDA Foreign Agriculture Service, Washington, DC.

USDA (2017c) Oil crops yearbook. USDA Economic Research Service. Available at: www.ers.usda.gov/data-products/oil-crops-yearbook.aspx (accessed 22 February 2017).

USDA (2017d) PSD online. USDA Foreign Agriculture Service. Available at: https://apps.fas.usda.gov/psdonline/app/index.html#/app/advQuery (accessed 23 February 2017).

USDA (2018) Quick stats. USDA National Agricultural Statistics Service. Available at: http://quickstats.nass.usda.gov/ (accessed 26 March 2018).

USSEC (2015) *Chapter One: the Soybean, its History, and its Opportunities*. United States Soybean Export Council, Chesterfield, Missouri. Available at: http://ussec.org/wp-content/uploads/2012/08/Chap1.pdf (accessed 15 January 2015).

Valent (2016) INTEGO™ SUITE soybeans. Available at: www.valent.com/agriculture/products/intego-suite-soybeans/index.cfm (accessed 15 June 2016).

van Bruggen, A.H.C. and Semenov, A.M. (2000) In search of biological indicators for soil health and disease suppression. *Applied Soil Ecology* 15, 13–24. https://doi.org/10.1016/S0929-1393(00)00068-8

van Bruggen, A.H.C. and Termorskuizen, A.J. (2003) Integrated approaches to root disease management in organic farming systems. *Australasian Plant Pathology* 32, 141–156. https://doi.org/10.1071/ap03029

van de Wiel, C., Smulders, M., Visser, R., and Schaart, J. (2016) *New Developments in Green Biotechnology – an Inventory for RIVM*. Wageningen UR Plant Breeding, Wageningen, The Netherlands. Available at: http://library.wur.nl/WebQuery/wurpubs/fulltext/385481 (accessed 24 September 2018).

van Es, H.M., Woodard, J.D., Glos, M., Chiu, L.V., Dutta, T., and Ristow, A. (2016) *Digital Agriculture in New York State: Report and Recommendations*. Cornell University, Ithaca, New York.

van Evert, F., Fountas, S., Jakovetic, D., Crnojevic, V., Travlos, I., and Kempenaar, C. (2017) Big Data for weed control and crop protection. *Weed Research* 57, 218–233. https://doi.org/10.1111/wre.12255

Vincelli, P. (2016) Genetic engineering and sustainable crop disease management: opportunities for case-by-case decision-making. *Sustainability* 8, 495–516. https://doi.org/10.3390/su8050495

Vleeshouwers, V.G., Rietman, H., Krenek, P., Champouret, N., Young, C., Oh, S.-K., Wang, M, Bouwmeester, K., Vosman, B., Visser, R.G., *et al.* (2008) Effector genomics accelerates discovery and functional profiling of potato disease resistance and *Phytophthora infestans* avirulence genes. *PLoS One* 3, e2875. https://doi.org/10.1371/journal.pone.0002875

Voytas, D.F. and Gao, C. (2014) Precision genome engineering and agriculture: opportunities and regulatory challenges. *PLoS Biology* 12, 1–6. https://doi.org/10.1371/journal.pbio.1001877

Walls, J.T., Caciagli, P., Tooker, J.F., Russo, J.M., Rajotte, E.G., and Rosa, C. (2016) Modeling the decision process for barley yellow dwarf management. *Computers and Electronics in Agriculture* 127, 775–786. https://doi.org/10.1016/j.compag.2016.08.005

Walter, M.W. (2002) Structure-based design of agrochemicals. *Natural Product Reports* 19, 278–291. https://doi.org/10.1039/b100919m

Wang, F., Wang, C., Liu, P., Lei, C., Hao, W., Gao, Y., Liu, Y.-G., and Zhao, K. (2016) Enhanced rice blast resistance by CRISPR/Cas9-targeted mutagenesis of the ERF transcription factor gene *OsERF922*. *PLoS One* 11, e0154027. https://doi.org/10.1371/journal.pone.0154027

Wang, Y., Cheng, X., Shan, Q., Zhang, Y., Liu, J., Gao, C., and Qiu, J.-L. (2014) Simultaneous editing of three homoeoalleles in hexaploid bread wheat confers heritable resistance to powdery mildew. *Nature Biotechnology* 32, 947–951. https://doi.org/10.1038/nbt.2969

Warnken, P.F. (1999) *The Development and Growth of the Soybean Industry in Brazil*. Iowa State University Press, Ames, Iowa.

Waterfield, G. and Zilberman, D. (2012) Pest management in food systems: an economic perspective. *Annual Review of Environment and Resources* 37, 223–245. https://doi.org/10.1146/annurev-environ-040911-105628

Weersink, A., Deen, W., and Weaver, S. (1991) Defining and measuring economic threshold levels. *Canadian Journal of Agricultural Economics* 39, 619–625. https://doi.org/10.1111/j.1744-7976.1991.tb03613.x

Weiland, J.E., Santamaria, L., and Grünwald, N.J. (2014) Sensitivity of *Pythium irregulare*, *P. sylvaticum*, and *P. ultimum* from forest nurseries to mefenoxam and fosetyl-Al, and control of Pythium damping-off. *Plant Disease* 98, 937–942. https://doi.org/10.1094/PDIS-09-13-0998-RE

Whalon, M.E., Mota-Sanchez, D., and Hollingworth, R.M. (2008) Analysis of global pesticide resistance in arthropods. In: Whalon, M.E., Mota-Sanchez, D., and Hollingworth, R.M. (eds) *Global Pesticide Resistance in Arthropods*. CAB International, Wallingford, UK, pp. 5–31.

White, J., Stanghellini, M., and Ayoubi, L. (1988) Variation in the sensitivity to metalaxyl of *Pythium* spp. isolated from carrot and other sources. *Annals of Applied Biology* 113, 269–277. https://doi.org/10.1111/j.1744-7348.1988.tb03303.x

Witcombe, J.R. and Virk, D.S. (2001) Number of crosses and population size for participatory and classical plant breeding. *Euphytica* 122, 451–462. https://doi.org/10.1023/a:1017524122821

Wohlgenant, M.K. (1997) The nature of the research-induced supply shift. *Australian Journal of Agricultural and Resource Economics* 41, 385–400. https://doi.org/10.1111/1467-8489.t01-1-00019

Wooldridge, J.M. (2009) *Introductory Econometrics: a Modern Approach*. South-Western Cengage Learning, Mason, Ohio.

Workneh, F., Tylka, G., Yang, X., Faghihi, J., and Ferris, J. (1999) Regional assessment of soybean brown stem rot, *Phytophthora sojae*, and *Heterodera glycines* using area-frame sampling: prevalence and effects of tillage. *Phytopathology* 89, 204–211. https://doi.org/10.1094/PHYTO.1999.89.3.204

Wrather, J.A. and Koenning, S.R. (2006) Estimates of disease effects on soybean yields in the United States 2003 to 2005. *Journal of Nematology* 38, 173–180.

Wrather, J.A. and Koenning, S.R. (2009) Effects of diseases on soybean yields in the United States 1996 to 2007. *Plant Health Progress*, https://doi.org/10.1094/PHP-2009-0401-1001-RS

Wrather, J.A., Chambers, A., Fox, J., Moore, W., and Sciumbato, G. (1995) Soybean disease loss estimates for the southern United States, 1974 to 1994. *Plant Disease* 79, 1076–1099.

Wrather, J.A., Anderson, T., Arsyad, D., Gai, J., Ploper, L., Porta-Puglia, A., Ram, H.H., and Yorinori, J. (1997) Soybean disease loss estimates for the top 10 soybean producing countries in 1994. *Plant Disease* 81, 107–110. https://doi.org/10.1094/PDIS.1997.81.1.107

Wrather, J.A., Anderson, T., Arsyad, D., Tan, Y., Ploper, L., Porta-Puglia, A., Ram, H.H., and Yorinori, J. (2001a) Soybean disease loss estimates for the top ten soybean-producing countries in 1998. *Canadian Journal of Plant Pathology* 23, 115–121. https://doi.org/10.1080/07060660109506918

Wrather, J.A., Stienstra, W., and Koenning, S.R. (2001b) Soybean disease loss estimates for the United States from 1996 to 1998. *Canadian Journal of Plant Pathology* 23, 122–131. https://doi.org/10.1080/07060660109506919

Wrather, J.A., Koenning, S.R., and Anderson, T. (2003) Effect of diseases on soybean yields in the United States and Ontario (1999–2002) *Plant Health Progress* 4, 1–24. https://doi.org/10.1094/PHP-2003-0325-01-RV

Wrather, J.A., Shannon, G., Balardin, R., Carregal, L., Escobar, R., Gupta, G., Ma. Z., Morel, W., Ploper, D., and Tenuta, A. (2010) Effect of diseases on soybean yield in the top eight producing countries in 2006. *Plant Health Progress*, https://doi.org/10.1094/PHP-2010-0125-01-RS

Wu, X.-L., Zhang, B.-Q., Sun, S., Zhao, J.-M., Yang, F., Guo, N., Gai, J.-Y., and Xing, H. (2011) Identification, genetic analysis and mapping of resistance to *Phytophthora sojae* of Pm28 in soybean. *Agricultural Sciences in China* 10, 1506–1511. https://doi.org/10.1016/S1671-2927(11)60145-4

Yang, X.B. (1999a) *Pythium* damping-off and root rot. In: Hartman, G.L., Sinclair, J.B., and Rupe, J.C. (eds) *Compendium of Soybean Diseases*, 4th edn. American Phytopathological Society, St. Paul, Minnesota, pp. 42–44.

Yang, X.B. (1999b) Rhizoctonia damping-off and root rot. In: Hartman, G.L., Sinclair, J.B., and Rupe, J.C. (eds) *Compendium of Soybean Diseases*, 4th edn. American Phytopathological Society, St. Paul, Minnesota, pp. 45–46.

Yang, X.B., Ruff, R., Meng, X., and Workneh, F. (1996) Races of *Phytophthora sojae* in Iowa soybean fields. *Plant Disease* 80, 1418–1420. https://doi.org/10.1094/PD-80-1418

Yoe, C. (2011) *Principles of Risk Analysis: Decision Making Under Uncertainty*. CRC Press, Boca Raton, Florida.

Yuen, J.E. and Hughes, G. (2002) Bayesian analysis of plant disease prediction. *Plant Pathology* 51, 407–412. https://doi.org/10.1046/j.0032-0862.2002.00741.x

Zapata, S.D., and Carpio, C.E. (2014) The theoretical structure of producer willingness to pay estimates. *Agricultural Economics* 45, 613–623. https://doi.org/10.1111/agec.12110

Zhang, B. and Yang, X.B. (2000) Pathogenicity of *Pythium* populations from corn–soybean rotation fields. *Plant Disease* 84, 94–99. https://doi.org/10.1094/PDIS.2000.84.1.94

Zhang, C., Wang, X., Zhang, F., Dong, L., Wu, J., Cheng, Q., Qi, D., Yan, X., Jiang, L., Fan, S., *et al.* (2017) Phenylalanine ammonia-lyase2.1 contributes to the soybean response towards *Phytophthora sojae* infection. *Scientific Reports* 7, 7242. https://doi.org/10.1038/s41598-017-07832-2

Zhang, S. (2011) Computer-aided drug discovery and development. In: Satyanarayanajois, S.D. (ed.) *Drug Design and Discovery: Methods and Protocols*. Humana Press, Totowa, New Jersey, pp. 23–38.

Zhang, S. and Xue, A.G. (2010) Population biology and management strategies of *Phytophthora sojae* causing *Phytophthora* root and stem rots of soybean. In: Arya, A. and Perello, A.E. (eds) *Management of Fungal Plant Pathogens*. CAB International, Wallingford, UK, pp. 318–328.

Zhao, X., Griffiths, W.E., Griffith, G.R., and Mullen, J.D. (2000) Probability distributions for economic surplus changes: the case of technical change in the Australian wool industry. *Australian Journal of Agricultural and Resource Economics* 44, 83–106. https://doi.org/10.1111/1467-8489.00100

Zhao, Y., Chang, X., Qi, D., Dong, L., Wang, G., Fan, S., Jiang, L., Cheng, Q., Chen, X., Han, D., *et al.* (2017) A novel soybean ERF transcription factor, *GmERF113*, increases resistance to *Phytophthora sojae* infection in soybean. *Frontiers in Plant Science* 8, 299. https://doi.org/10.3389/fpls.2017.00299

Zimdahl, R.L. (2007) *Fundamentals of Weed Science*. Academic Press, Burlington, Massachusetts.

Index

Note: Page numbers in **bold** type refer to **figures**
Page numbers in *italic* type refer to *tables*
Page numbers followed by 'n' refer to notes